FOR THE
IB DIPLOMA

Internal Assessment for
Biology

Skills for success

Andrew Davis

HODDER
EDUCATION

Dedication

For Stephen Sutton and John Grahame, who taught me ecology at Leeds and were instrumental to subsequent success.

Author profile

Andrew Davis teaches IB Diploma Biology and Environmental Systems and Societies (ESS) at St Edward's School, Oxford. He is the author of several IB textbooks, including *Biology for the IB Diploma Study and Revision Guide* and *Biology for the MYP 4 & 5: By concept.* He is also author of online teaching and learning resources: *Biology for the IB Diploma Teaching and Learning* and *Biology for the IB MYP 4 & 5 Dynamic Learning.*

Author's acknowledgments

Christopher Talbot, formerly of Anglo-Chinese School (Independent) Singapore, was instrumental in formulating the style and content of this book, and freely provided a substantial amount of material and input, including rigorous editing, that significantly improved the text – I am extremely grateful to him for providing such tremendous support and encouragement throughout the process. My thanks also to Dr David Fairley of Overseas Family School, Singapore, who provided invaluable feedback on all chapters of the book – his input is greatly appreciated. Dr Robert L. Dean, The University of Western Ontario, Canada, and Dr Jon Nield, Queen Mary University of London, also generously gave of their time and expertise.

Photo credits

The Publishers would like to thank the following for permission to reproduce copyright material.

pp.vi–vii © nevodka/123rf; **p.8** Rhinovirus © fotoliaxrender/Adobe Stock; Measles virus © Kateryna Kon/Shutterstock; *E. coli* © abhijith3747/Adobe Stock; Pollen grain © tonaquatic/Adobe Stock; *Paramecium* © sinhyu/Adobe Stock; Sesame seed © Teodora_D/Adobe Stock; Woodlouse © Anton/Adobe Stock; Meerkat © EcoView/Adobe Stock; Human © PIC4U/Adobe Stock; Sperm whale © wildestanimal/Adobe Stock; Redwood tree © nstanev/Adobe Stock; Seagrass colony © trekandphoto/Adobe Stock; Great barrier reef © superjoseph/Adobe Stock; **p.13** © Andrew Davis; **p.41** © Andrew Davis; **p.42** *r* © sudok1/Adobe Stock; **p.50** *t, b* © sinhyu/Adobe Stock; **p.72** *l, m* © Andrew Davis; **p.72** *r* © Garrett Nagle; **p.74** *r* © Andrew Davis; **p.79** *l, r* © Andrew Davis; **p.80** © Andrew Davis; **p.143** © bloomicon/Adobe Stock

Every effort has been made to trace all copyright holders, but if any have been inadvertently overlooked, the Publishers will be pleased to make the necessary arrangements at the first opportunity.

Orders: please contact Hachette UK Distribution, Hely Hutchinson Centre, Milton Road, Didcot, Oxfordshire, OX11 7HH. Telephone: +44 (0)1235 827827. Email education@hachette.co.uk Lines are open from 9 a.m. to 5 p.m., Monday to Friday. You can also order through our website: www.hoddereducation.com

© Andrew Davis 2018

First published in 2018 by

Hodder Education,

An Hachette UK Company

Carmelite House

50 Victoria Embankment

London EC4Y 0DZ

www.hoddereducation.com

Impression number 10 9 8 7 6 5 4 3

Year   2022

Cover photo © Patryk Kosmider – Fotolia

Illustrations by Aptara Inc.

Typeset in Goudy Oldstyle Std 10/12 by Aptara Inc.

Printed and bound by CPI Group (UK) Ltd, Croydon, CR0 4YY

A catalogue record for this title is available from the British Library.

Contents

Introduction

How to use this book

There are two aspects to the practical work in the IB Biology programme: general practical work and a single individual investigation – the internal assessment project.

This book has been written for IB Biology students and is to be used throughout your two years of study and to support your preparation for the internal assessment.

Practical activities and the IA form an essential part of the 2014 syllabus (first assessment held in 2016), making up 40 hours of recommended teaching time for SL and 60 hours for HL. This represents an average 25 % of the total teaching time. The IA is worth 20 % of the final assessment.

General practical work includes experiments, ecological fieldwork studies, spreadsheet or online simulations, demonstrations by your biology teacher and class activities which will be of a formative nature. These are designed to help you learn biology via practical work.

The 'Applications and skills' section of the syllabus lists specific lab skills, techniques and experiments that you must experience at some point during your study of the IB Biology course. Your school is likely to arrange additional practical work covering other topics in the IB Biology programme. Please note it is the skills and not the specific experiments that will be assessed in the written examinations. Other recommended lab skills, techniques and experiments are listed in the 'aims' section of the IB Biology syllabus pages.

Within the IB Biology syllabus, there is also a specific set of mandatory practicals that you will carry out over the course and your knowledge and understanding of these practicals will be assessed in your final examination papers.

This guide will ensure you can aim for your best grade by:

- building practical and analytical skills for the mandatory and other common practicals through a comprehensive range of strategies and detailed examiner advice and expert tips

- offering concise, clear explanations of all the IB requirements, such as the assessment objectives of each assessment criterion for the IA, including check lists and rules on academic honesty

- demonstrating what is required to obtain the best IA grade for the Individual Investigation with advice and tips, including common mistakes to avoid

- suggesting practicals that might, if modified, form the basis of an Individual Investigation

- making explicit reference to the IB learner profile and the associated Approaches to Learning (ATL) that are central to the IB programme, with their connections to practical work

- providing infographics with every opening chapter which visually display essential information

- including exemplars and worked answers and commentary throughout so you can see the application of biological principles and concepts

- testing your comprehension of the skills covered with embedded activity questions.

Use the space provided in the margins of the book to make your own notes and record your own observations as you progress through the course.

Practical skills required for Option A and Option B are covered at the end of the book. Skills needed for Option C (Ecology) and Option D (Physiology) are contained in Chapters 4 and 5.

Features of this book

Key definition

The definitions of essential key terms are provided on the page where they appear. These are words that you are expected to know for exams and practical work. A glossary of other essential terms, highlighted throughout the text, is given at the end of the book.

Examiner guidance

These tips give you advice that is likely to be in line with IB examiners.

Worked examples

Some practical skills require you to carry out mathematical calculations, plot graphs, and so on. These examples show you how.

■ ACTIVITY

Suggested outline of possible practice activities with the command terms highlighted.

Ideas for investigations

Ideas for possible investigations.

Expert tip

These tips give practical advice that will help you boost your final grade.

Common mistake

These identify typical mistakes that candidates make and explain how you can avoid them.

IB Learner Profile

The IB Biology course is linked to the IB learner profile. Throughout the course, and while carrying out your internal assessment, you will have the opportunity to develop each aspect of the learner profile: Inquirers, Knowledgeable, Thinkers, Communicators, Principled, Open-minded, Caring, Risk-takers, Balanced and Reflective.

Studying IB Biology

Practicals

Carrying out practicals throughout your IB Biology course will give you the opportunity to practice carrying out an investigation, and will give you the scientific skills you need for your internal assessment.

Biology

Approaches to Learning

The IB Biology course, and the internal assessment in particular, give you the chance to develop the approaches to learning skills:

- thinking skills when planning investigations, collecting data and analyzing your results
- social skills when working with your peers
- communication skills when reporting and presenting your findings
- self-management skills when working independently
- research skills to help plan your investigation, and to put it into context.

Internal Assessment

The internal assessment gives you the opportunity to display the skills and knowledge you have learned throughout your course, while exploring an area of Biology that interests you personally.

Studying IB Biology

Biology and the scientific method

Biology is both an observational and experimental science. Observations about the natural world result in hypotheses, which can lead to **investigations** that manipulate a variable in order to see its effect, which in turn result in an improved understanding of biology. This **scientific method** can be seen as a cycle (Figure 1 and Table 1). Exploration of one idea can lead to further modification, through reflection and evaluation, resulting in the investigation of further hypotheses.

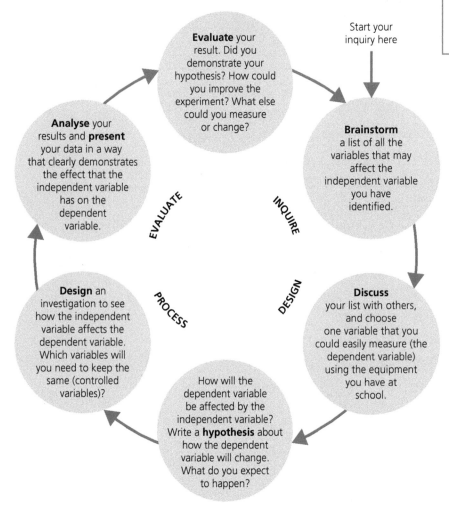

Figure 1 The investigation cycle

Stage of cycle	Description	Key word definitions
1	Formulate your research question: which usually inquires how one variable (the independent variable) affects another (the dependent variable).	Variable – a factor that is changed, measured, or kept the same in an investigation. Independent variable – the variable that is changed in an investigation. Dependent variable – the variable that is measured in an investigation. Processed variable – a variable that can be produced by transforming a measured variable through mathematical manipulation (see Expert tip box on page x).

Stage of cycle	Description	Key word definitions
2	The research question may lead you to formulate a hypothesis. When making a hypothesis, the investigator proposes how the independent variable may affect the dependent variable. A prediction may be made.	Hypothesis – a tentative explanation of an observed phenomenon or event that can be investigated using the scientific method. Prediction/predict – give an expected result.
3	So that a correlation between the two variables (independent and dependent) can be established, other variables must be kept the same – these are called controlled variables.	Controlled variable – a variable that is kept the same in an investigation. In an experiment, at least three controlled variables should be listed, and information about how they will be kept the same included.
4	Develop a method and outline it clearly. Materials should be listed. Sizes, volumes and other appropriate information should also be included, such as ± measurement uncertainties (see page 129). One variable should be manipulated and one measured; all other variables should be controlled. The method should be written in the passive voice and in sufficient detail and clarity so that someone else can follow the instructions.	Control – an experiment where the independent variable is either kept constant or removed. This can be used for comparison, to prove that any changes in the dependent variable in experiments when the independent variable is manipulated must be due to the independent variable rather than other factors.
5	Carry out your investigation and gather data. Record your data by measuring the dependent variable. Present your data with their appropriate units. Process your data in some way, for example, mean values calculated or a line of best fit drawn. Plot a graph to present the results in a way that displays them clearly and helps interpretation. Qualitative data (observations) may also be recorded (see Expert tip box on page x).	Data – recorded products of observations and measurements. Measuring/measure – obtain a value for a quantity.
6	Following an analysis, develop an explanation for the results. What do the results show? Describe and explain the results. Do the results support the hypothesis, or not?	Analysis/analyse – break down in order to bring out the essential elements or structure. Explanation/explain – give a detailed account including reasons or causes.
7	Evaluate the investigation and suggest improvements. When commenting on limitations, consider the procedures, the equipment, the use of equipment, the quality of the data (for example, its accuracy and precision) and the relevance of the data. To what extent may the limitations have affected the results? Propose realistic improvements that address the limitations. The sensitivity of the equipment used must also be taken into account.	Evaluated/evaluation – make an appraisal by weighing up the strengths and limitations. Accuracy – how close to the true value a result is. Precision – describes the reproducibility of repeated measurements of the same quantity and how close they are to each other. Note, measurements can be precise but not accurate (see Figure 2). Sensitivity – the number of significant digits to which a value can be reliably measured. For example, if a digital thermometer can measure to two decimal places, this is the sensitivity of data that can be recorded.
8	The improvements to the method can lead to further investigations, and so the cycle repeats itself.	

Table 1 The scientific method cycle

Expert tip

Controls can be used in biology to prove that changes in the dependent variable are due to manipulation of the independent variable, rather than other factors. The results of the investigation are compared with the results of the control. For example, control groups can be used in physiology experiments to provide a baseline measure: members of this group are identical with all other subjects being investigated but they do not receive the treatment or experimental manipulation that the treatment group receives.

Key definition
Treatments – well-defined conditions applied to the sample units.

Replicates can improve the reliability of an investigation and enable anomalies to be identified.

Due to the complexities of biological systems, other variables besides the independent variable may affect the dependent variable. These **confounding variables** must be held constant if possible, or at least monitored so that their effect on the results can be accounted for in the analysis.

In cases where confounding variables cannot easily be controlled, blocks of experimental and control groups can be distributed so that the influence of any confounding variable is likely to be the same across the experimental and control groups.

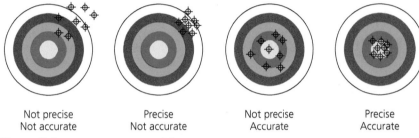

| Not precise | Precise | Not precise | Precise |
| Not accurate | Not accurate | Accurate | Accurate |

Figure 2 Accuracy versus precision

Expert tip

Data can be either quantitative or qualitative. Qualitative data are observations not involving measurements, such as those recorded in an ecological study to note conditions in a survey area (for example, climatic conditions; factors that might make the collection site typical or atypical). Quantitative data are numerical data from measurements, such as measurements taken when recording a dependent variable (such as the rate of an enzyme reaction).

Quantitative data are numerical data (with associated units). An example of such data might be measurements of the breadth of leaves from a **species** of plant grown in shaded and exposed positions, or the pH values of top soil samples in different positions within tropical primary rainforest.

Quantitative data can be discontinuous (discrete) or continuous. Examples of discontinuous data are the number of petals present in a species of flower or the number of fruit flies on an apple. Examples of continuous data might be the masses of the individuals of a **population**, or the time taken for pea seeds to germinate after treatment with different concentrations of sulfur dioxide gas.

Expert tip

Raw data are data collected without any processing. They are simply the values of each variable collected. Raw data are often difficult to use for data analysis, and usually need to be processed in some way.

Processed data are data that are ready for analysis. Processing can include merging, sub-setting, or transforming data (carrying out a calculation).

A processed variable is a variable that is calculated from measured data (that is, from the dependent variable). For example, in enzyme practicals the dependent variable might be the volume of **product** formed per minute and the processed variable calculated from this could be the **rate of reaction**.

During your two-year IB Biology course you will carry out a significant amount of practical work. This work may be lab-based, or located in an environment outside your school or college. This work will help you to understand some of the biological concepts, teach you important practical and analytical skills, and give you opportunities to extend your knowledge through your own investigations.

Examiner guidance

Be aware that a strong correlation does not necessarily mean causation. In other words, you cannot state definitely that X causes the changes in Y (where X is a dependent variable and Y is an independent variable).

You may of course suggest in your conclusion that X causes changes in Y, or *vice versa*. However, you should not draw definite cause and effect conclusions based on correlation.

There are several reasons why you cannot make definite causal statements:

- You do not know the direction of the cause – does X cause Y, or does Y cause X?
- A third variable Z might be involved that is responsible for the correlation between X and Y.
- The apparent relationship might simply be due to chance.

Framework for Biology

Your IB Biology syllabus is comprehensive and detailed. It helps to simplify the content by using a 'concept tree' (Figure 3), which outlines the essential components of your course and how they interrelate. The practical skills you will be learning during the course can be framed in the context of this concept tree. Subsequent chapters will cover essential skills, subdivided by the different levels of integration shown in Figure 3, that is, biochemistry, cell structure and function, physiology, and ecology.

Figure 3 can be used to help you decide the area of biology you want to address in your IA project (Chapter 8).

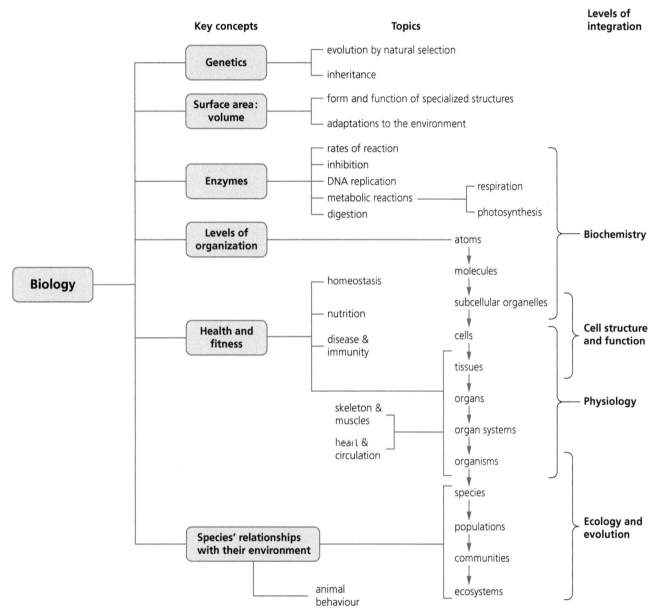

Figure 3 Biology concept tree

IB Biology practicals

There are numerous practicals listed in the *IB Biology Guide*. This book focuses on practical experiments that can help you not only with exam questions (Paper 3, section A) but also in the selection and implementation of a suitable investigation for your IA.

The practicals in the guide can be divided into four categories:

- Mandatory practicals: these are prescribed in the *IB Biology Guide*. You need an understanding of these experiments as they may be examined in Paper 3.

- Further practical skills: skills associated with these practicals may be assessed in Paper 3.

- Suggested practicals: useful additional practicals suggested in the *IB Biology Guide*, to be chosen at the discretion of your teacher. These will not specifically be examined, but might provide useful ideas to help you select and then implement your IA project.

- Computer simulations: information communication technology (ICT) is encouraged throughout all aspects of your course. Certain skills involving ICT are specified in the *IB Biology Guide*: these use computers to model or draw associations between complex data.

Mandatory practicals

Subtopic	Mandatory practical
1.1 Introduction to cells	Practical 1 (page 42) Use of a light microscope to investigate the structure of cells and tissues, with drawing of cells Calculation of the magnification of drawings and the actual size of structures and ultrastructures shown in drawings or micrographs
1.4 Membrane transport	Practical 2 (page 46) Estimation of osmolarity in tissues by bathing samples in hypotonic and hypertonic solutions
2.5 Enzymes	Practical 3 (page 17) Experimental investigation of a factor affecting enzyme activity
2.9 Photosynthesis	Practical 4 (page 28) Separation of photosynthetic pigments by chromatograph
4.1 Species, communities and ecosystems	Practical 5 (page 80) Setting up sealed mesocosms to try to establish sustainability
6.4 Gas exchange	Practical 6 (page 51) Monitoring of ventilation in humans at rest and after mild and vigorous exercise
9.1 (*AHL*) Transport in the xylem of plants	Practical 7 (page 53) Measurement of transpiration rates using potometers

Table 2 List of mandatory practicals for Biology

Further practicals skills

The practicals in Table 3 are listed as 'skills' in the IB Biology Guide.

Subtopic	Practical skill
2.5 Enzymes	Design of experiments to test the effect of temperature, pH and substrate concentration on the activity of enzymes
2.8 Cell respiration	Analysis of results from experiments involving measurement of respiration rates in germinating seeds or invertebrates using a respirometer
2.9 Photosynthesis	Design of experiments to investigate the effect of limiting factors on photosynthesis
3.5 Genetic modification and biotechnology	Design of an experiment to assess one factor affecting the rooting of stem-cuttings
4.1 Species, communities and ecosystems	Testing for association between two species using the chi-squared test with data obtained by quadrat sampling Recognizing and interpreting statistical significance
8.1 Metabolism	Calculating and plotting rates of reaction from raw experimental results
9.1 Transport in the xylem of plants	Design of an experiment to test hypotheses about the effect of temperature or humidity on transpiration rates
9.2 Transport in the phloem of plants	Analysis of data from experiments measuring phloem transport rates using aphid stylets and radioactively labelled carbon dioxide
9.4 Reproduction in plants	Design of experiments to test hypotheses about factors affecting germination

Table 3 Practical skills

▦ Suggested practicals

The following suggested practicals will help you enhance your understanding of biology. Your teacher might select them to include in your Practical Scheme of Work (PSOW). They are listed in the 'guidance' section of the syllabus and so will not be examined.

Subtopic	Suggested practical
1.4 Membrane transport	Dialysis tubing experiments can act as a model of membrane action
	Experiments with potato, beetroot or single-celled algae can be used to investigate real membranes
2.4 Proteins	Egg white or albumen solutions can be used in thermal denaturation experiments
2.5 Enzymes	Lactase can be immobilized in alginate beads and experiments can then be carried out in which the lactose in milk is hydrolyzed
8.1 Metabolism	Experiments on enzyme inhibition
8.3 Photosynthesis	Hill's method demonstrating electron transfer in chloroplasts by observing DCPIP reduction
	Immobilization of a culture of an alga such as *Scenedesmus* in alginate beads
	Measurement of the rate of photosynthesis by monitoring the effect on hydrogencarbonate indicator
9.1 Transport in the xylem of plants	Measurement of stomatal apertures using leaf casts, including replicate measurements to enhance reliability
	Measurement of the distribution of stomata using leaf casts, including replicate measurements to enhance reliability

Table 4 Suggested practicals

▦ Computer simulations

Other practical skills involve the use of ICT (Table 5).

Subtopic	Activity/simulation
2.3 Carbohydrates and lipids	Use of molecular visualization software to compare cellulose, starch and glycogen
3.1 Genes	Use of a database to determine differences in the base sequence of a gene in two species
3.2 Chromosomes	Use of databases to identify the locus of a human gene and its polypeptide product
7.1 DNA structure and replication	Use of molecular visualization software to analyse the association between protein and DNA within a nucleosome
7.3 Translation	Use of molecular visualization software to analyse the structure of eukaryotic ribosomes and a tRNA molecule

Table 5 Computer activities

Approaches to learning

Approaches to learning (ATLs) are deliberate strategies, skills and attitudes that underlie all aspects of the IB Diploma Programme. These approaches are intrinsically linked with the IB learner profile attributes (see below), and are designed to enhance your learning and preparation for the Diploma Programme assessment and beyond.

The aims of ATLs in the IB Diploma Programme are to:

- link prior knowledge to course-specific understandings, and make connections between different subjects

- encourage you to develop a variety of skills that will equip you to continue to be actively engaged in learning after you leave your school or college

- help you not only to obtain university admission through better grades but also to prepare for success during tertiary education and beyond

- enhance the coherence and relevance of your IB Diploma Programme experience.

The five approaches to learning develop the following skills:

- thinking skills
- social skills

Expert tip

ATLs encompass the key values and principles that underpin an IB education.

■ communication skills

■ self-management skills

■ research skills.

Practical activities clearly allow you to interact directly with natural phenomena, explore a topic and examine specific research questions. All practical skills covered in this book can be viewed in the context of ATLs. They also give you the opportunity to develop and use IB terminology:

■ research skills to find out appropriate methods to investigate specific research questions, and put your investigation in the context of the wider scientific community

■ thinking skills to design investigations, collect and analyse data, and then evaluate your results

■ social skills in order to collaborate with peers

■ communication skills to effectively and concisely present your findings

■ self-management skills to make sure you successfully plan your time and meet deadlines.

The IB learner profile

The IB Biology course is closely linked with the IB learner profile (Table 6). By following the course, you will have engaged with all attributes of the IB learner profile: the requirements of the IA provide opportunities for you to develop every aspect of the profile.

Learner profile attribute	Relevance to Biology syllabus
Inquirers	Practical work and internal assessment
Knowledgeable	Links to international-mindedness
	Practical work and internal assessment
Thinkers	Links to theory of knowledge
	Practical work and internal assessment
Communicators	External assessment (examinations)
	Practical work and internal assessment
Principled	Practical work and internal assessment
	Ethical behaviour
	Academic honesty
Open-minded	Links to international-mindedness
	Practical work and internal assessment
	The group 4 project
Caring	Practical work and internal assessment
	The group 4 project
	Ethical behaviour
Risk-takers	Practical work and internal assessment
	The group 4 project
Balanced	Practical work and internal assessment
	The group 4 project
	Fieldwork
Reflective	Practical work and internal assessment
	The group 4 project

Table 6 Relevance of the IB learner profile to the IB Biology syllabus

The internal assessment

The internal assessment forms 20% of your final mark, with the external examinations (Papers 1, 2 and 3) forming 80% of your mark. The assessment and the assessment criteria are the same for standard level and higher level.

Criterion	Personal engagement	Exploration	Analysis	Evaluation	Communication	Total marks available
Marks available	2	6	6	6	4	24

Table 7 Marking criteria for the biology internal assessment

Your internal assessment mark is based upon one scientific investigation known as the Individual Investigation. This will involve 10 hours of work and you will generate a word-processed report or write-up 6 to 12 pages long.

This will be marked out of a maximum of 24 marks based upon the five group 4 assessment criteria (Table 7). Your mark out of 24 will then be scaled to a mark out of 20. Your Individual Investigation will be internally marked by your IB Biology teacher but moderated externally (re-marked) by an experienced IB Biology teacher appointed by the IBO.

There are separate chapters for each of the internal assessment criteria (Chapters 8–12). Checklists at the end of each criterion chapter will help you to ensure that your report matches the requirements of the group 4 assessment criteria.

Grade boundaries for the internal assessment are as follows (using data from May 2016–May 2017 examinations):

Grade	1	2	3	4	5	6	7
Mark range	0–3	4–6	7–10	11–13	14–16	17–19	20–24

Table 8 Grade boundaries for the biology internal assessment

Planning an internal assessment

There are no IB requirements in terms of planning, a time line or documentation, but your school might require you to complete a preliminary internal assessment proposal. For this, you might need to suggest a research question and methodology, carry out a risk assessment and complete a requisition for apparatus, biological materials and biochemicals (for example, enzymes) for preliminary work.

Setting up a schedule

It might be helpful to set up a timeline with start dates and deadlines for each part of your Individual Investigation, if your school has not done so. A sample timeline is shown in Table 9.

Stage in investigation	Start date	Task	Deadline date
Planning 1		Read Chapters 8 and 9 in this guide.	
Planning 2		Decide on the research question, methodology (including controls and statistical test, if appropriate) and outline data collection methods.	
Planning 3		Prepare a risk assessment for these experiments and show your biology teacher the completed Risk Assessment form.	
Planning 4		Ensure that the apparatus, biological materials and chemicals you need will be available in your school laboratory.	
Practical		Complete the experimental work and collect raw data in the time allocated. Allow time for preliminary work, and to carry out duplicates, extending the range of data collected. Document any alterations to your plan as soon as they occur and, if necessary, make alterations to the supporting theory.	
Report 1		Hand in the first draft and consult with your biology teacher.	
Report 2		Submit the final draft after an online plagiarism check.	

Table 9 An example internal assessment timeline

Experimental skills

SI units

Scientific notation

Orders of magnitude

Significant figures

Significant figures

Measurement

Mathematical and measurement skills

and abilities

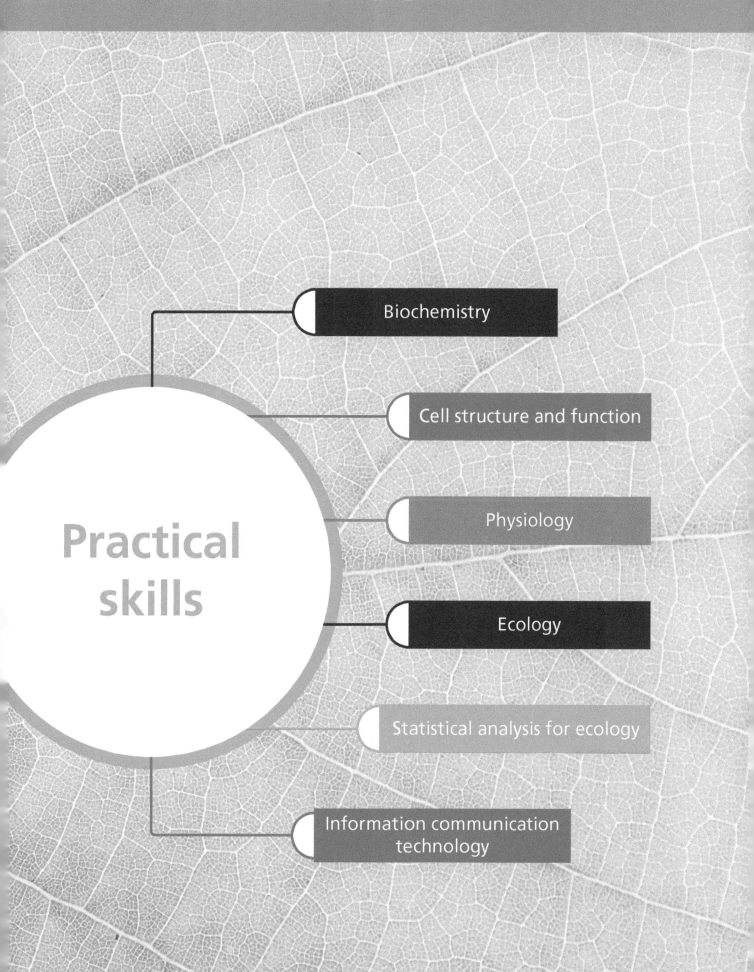

Biochemistry

Cell structure and function

Physiology

Ecology

Statistical analysis for ecology

Information communication technology

Practical skills

1 Mathematical and measurement skills

SI units

In the sciences, the metric system is used to represent physical units. Metric measurement is used worldwide to improve communication and to ensure that standard methods are employed. The metric system uses base units, such as metre (for length) and gram (for mass), which can be modified using prefixes, such as kilo- and milli- (see below), to reduce or enlarge the base units by factors of ten.

An international agreement was reached in 1960 that specified units for use in scientific measurements. These units are called SI units (after the French *Système International d'Unités*). Measurements in biology are usually recorded using SI units. The SI system has specified base units from which all other units are derived. The main units used in biology (quantity, unit, and SI unit symbol) are:

- **length** – metre (m)
- **mass** – kilogram (kg)
- **time** – second (s)
- **temperature** – degrees Celsius (°C)*
- **force** – newton (N)
- **pressure** – pascal (Pa)
- **energy** – joule (J)
- **volume** – litre (l)+
- **amount of a substance** – mole (mol).

* The kelvin (K) is the correct SI unit for temperature, but is rarely used in biology, where degrees Celsius is more common.

+ Volume should, strictly speaking, be measured in cubic metres, m^3, not litres ($1 l = 1 dm^3$), however litre is more useful in biology and so is widely used.

Derived SI units

The SI base units can be used to derive the units of other quantities. In each case, an equation is used to define the derived quantity, substituting the appropriate base units. For example, speed is calculated by dividing distance by time, and so the SI unit for speed is the SI unit for distance (metres) divided by the SI unit for time (seconds), that is, $m\ s^{-1}$ (metres per second). Other units of measurements include:

- **concentration** – $g\ l^{-1}$ (or $mol\ l^{-1}$)
- **rate of reaction** – $mol\ l^{-1}\ s^{-1}$, that is, change in concentration/per unit time (or any other measure of progress/any unit of time). If gases are involved, $cm^3\ s^{-1}$ or $dm^3\ s^{-1}$ may be used. If $g\ min^{-1}$ are measured (that is, change in mass over time) this is not strictly speaking the rate of reaction, but is **proportional** to the rate of reaction.

Some measurements, such as logarithmic functions and absorbance, have no units. The pH scale, for example, is logarithmic. A logarithmic scale compresses the range of values, giving more space to smaller values while reducing the space available for larger values. Each cycle on the scale increases by a power of 10: for example, in the first cycle, values would be 1, 2, 3, 4, etc., whereas in the second cycle they would be 10, 20, 30, 40, etc., and in the third cycle 100, 200, 300, 400, etc. See Chapter 10, page 128, for further information about logarithmic scales and how they are plotted.

See Chapter 10, page 128, for further information about logarithmic scales and how they are plotted.

Expert tip

Here are some general rules when applying SI units:

- Units are always spelled beginning with a lower-case letter, for example, metre. This is also the case when they are named after a scientist, for example, joule.
- Units are always expressed in the singular not the plural, for example, 2 min not 2 mins. (*NB: This is the case when using the unit symbol, but not when reading the quantity out loud, for example, 5kg is correct but this is read as 5 kilograms*).
- There should be a space between the value and its symbol, for example, 5.00 kg not 5.00kg.
- Use the negative exponent ($^{-1}$) when expressing units rather than using a forward slash, for example, $m\ s^{-1}$ not m/s.

Expert tip

Examinations usually use the unit cm^3 rather than l ($1\ cm^3 = 1\ ml$).

Common mistake

When referring to units of temperature, never use the term 'centigrade'. The correct unit of measurement is degrees Celsius.

Expert tip

Quantities that represent ratios of two values, such as absorbance, do not have units. Both pH and absorbance are unit-less since they are logarithmic functions (pH = $-log_{10}$ [H^+(aq)]; absorbance = $log_{10}(I_0/I)$), and logarithms are always pure numbers which have no units.

Expert tip

The fact that pH is logarithmic means that pH 6.0 is 10 times more acidic than pH 7.0; natural rainwater at pH 5.5 is about 25 times more acidic than distilled water at pH 7.0. Acid rain is frequently more than 20 times more acidic than natural rainwater.

Non-SI units

Some measurements in biology do not use SI units. Some of these have already been explored (such as volume measurement in litres and temperature measurement in degrees Celsius). Other non-SI units are shown in Table 1.1; some of these units are used in the IB Biology course.

Physical quantity	Non-SI unit	Unit symbol	Conversion factor/use
energy	calorie	cal	1 cal = 4.184 J
length	Ångstrom	Å	1 Å = 10^{-10} m
mass	tonne	t	1 t = 10^3 kg
light	lux	lx	1 lx = 1 lm m^{-2} (one lumen per square metre)
sedimentation coefficient	Svedberg unit	S	Used, for example, to compare ribosomes (prokaryotic ribosomes = 70 S, eukaryotic = 80 S); an object's mass, density and shape determine its S value
heart rate	beats per minute	bpm	Varies according to the body's physical needs
genome size	base pair	bp	kilobase pair (kbp) = 1,000 bp
cells per sample	colony forming unit	cfu	Used to estimate the number of viable bacteria or fungal cells in a sample
particles per volume	plaque-forming unit	pfu	Used to estimate the number of particles capable of forming plaques per unit volume, for example, virus particles

Table 1.1 Non-SI units

Blood pressure and gas pressure may be expressed in millimetres of mercury (mm Hg). This refers to the height of a column of mercury that can be supported by the gas pressure being measured. Atmospheric pressure at 25 °C = 760 mm Hg = 101 000 Pa = 101 kPa.

The energy content of dried foods or fuels is usually expressed in kilojoules or kilocalories. A calorie (C) is the amount of heat energy required to heat one gram of water through one degree Celsius. 1 calorie = 4.184 J.

■ **ACTIVITY**

1 The length of the C–N bond in the peptide bond in proteins is 0.133 nm. Express this in micrometres, picometres and Angstroms (Å); 1 Å = 10^{-10} m, a non-SI unit which is commonly used to describe protein structures (http://www.rcsb.org).

Scientific notation

Numbers in science are often extremely large or extremely small. Consider the mass of a tobacco mosaic virus and the mass of the Sun, for example. They can be written as: 0.000 000 000 000 000 000 000 000 068 kilograms and 1 989 100 000 000 000 000 000 000 000 000 kilograms, respectively.

However, this notation uses many zeros and so there is a possibility of making a mistake when writing a value. Scientific notation is a way of expressing large and small numbers while avoiding lots of zeros. It uses the form: $N \times 10^n$, where N is a number between 1 and 10 and n is the exponent or the power to which 10 is raised. So in scientific notation, the mass of a tobacco mosaic virus can be written as 6.8×10^{-26} kg and the mass of the Sun as 1.9891×10^{30} kg.

As the definition implies and the following examples show, any number – not just large or small numbers – can be expressed in scientific notation:

$97\,400 = 9.74 \times 10^4$

$106.8 = 1.068 \times 10^2$

$10 = 1 \times 10^1$

$0.0029 = 2.9 \times 10^{-3}$

$0.005810 = 5.810 \times 10^{-3}$

Scientific notation is not merely a more convenient way of expressing numbers, it makes it easier to track significant figures.

Standard prefixes can be used to record large or very small numbers:

For large numbers:

10^3 = kilo (k) – for example, kilometre (km)

10^6 = mega (M) – for example, megametre (Mm)

10^9 = giga (G) – for example, gigametre (Gm)

10^{12} = tera (T) – for example, teragram (Tg)

For small numbers:

10^{-3} = milli (m) – for example, millilitre (ml)

10^{-6} = micro (μ) – for example, micrometre (μm)

10^{-9} = nano (n) – for example, nanometre (nm)

10^{-12} = pico (p) – for example, picometre (pm)

Figure 1.1 shows the importance of prefixes when referring to different levels of scale.

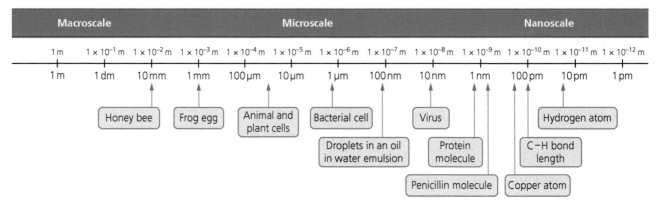

Figure 1.1 Units of distance from large scale (macroscale) to very small scale (nanoscale)

■ **ACTIVITIES**

2 Write the following numbers in scientific notation: 1 002, 54, 6 926 300 000, −393, 0.00361 and −0.0038.

3 Write the following numbers in ordinary notation: 1.93×10^3, 3.052×10^1, -4.29×10^2, 6.261×10^6 and 9.513×10^{-8}

4 What is the name given to the unit that equals (a) 10^{-9} gram; (b) 10^{-6} second; (c) 10^{-3} metre?

5 (a) What decimal fraction of a second is a picosecond, ps? (b) Express the measurement 4.0×10^3 m using a prefix to replace the power of ten. (c) Use standard exponential notation to express 4.56 mg in grams.

Orders of magnitude

Biologists have to be able to ascertain and record very large and very small measurements. Biological objects vary from tens of nanometres, such as viruses (for example, the Porcine circovirus type 1 has a diameter of 17 nm), to tens or hundreds of metres. The largest mammal is the blue whale (*Balaenoptera musculus*) which can measure up to 30 metres in length. The largest living organism on Earth is believed to be a honey fungus (*Armillaria ostoyae*), found in the Blue Mountains in Oregon, USA, measuring 3.8 km across.

The largest organisms are clonal colonies (made of many individual organisms) which can spread over large areas. For example, a seagrass colony (*Posidonia oceanica*) found in the Mediterranean sea may measure 8 kilometres across. The largest biological structure is the Great Barrier Reef in Queensland, Australia, which stretches 2 000 km and is composed of many individual organisms.

Scientists describe the magnitude or size of numbers using 'orders of magnitude'. An order of magnitude can be thought of as the power of ten closest to a given quantity. This allows scientists to quickly estimate the size of an object, or the difference in measure between two biological objects, by expressing it as a power of ten. Table 1.2 shows the orders of magnitude on the measurement of length.

Unit	Symbol	Conversion to standard unit (m)		
picometre	pm		1 pm =	1×10^{-12} m
			10 pm =	1×10^{-11} m
			100 pm =	1×10^{-10} m
nanometre	nm	1 000 pm =	1 nm =	1×10^{-9} m
			10 nm =	1×10^{-8} m
			100 nm =	1×10^{-7} m
micrometre	μm	1 000 nm =	1 μm =	1×10^{-6} m
			10 μm =	1×10^{-5} m
			100 μm =	1×10^{-4} m
millimetre	mm	1 000 μm =	1 mm =	1×10^{-3} m
			10 mm =	1×10^{-2} m
			100 mm =	1×10^{-1} m
metre	m	1 000 mm =	1 m =	1×10^{0} m
			10 m =	1×10^{1} m
			100 m =	1×10^{2} m
kilometre	km	1 000 m =	1 km =	1×10^{3} m
			10 km =	1×10^{4} m
			100 km =	1×10^{5} m
megametre	Mm	1 000 km =	1 Mm =	1×10^{6} m
			10 Mm =	1×10^{7} m
			100 Mm =	1×10^{8} m
gigametre	Gm	1 000 Mm =	1 Gm =	1×10^{9} m
			10 Gm =	1×10^{10} m
			100 Gm =	1×10^{11} m
terametre	Tm	1 000 Gm =	1 Tm =	1×10^{12} m

Table 1.2 Orders of magnitude and standard units

Expert tip

- One order of magnitude represents a ten-fold difference.
- Scientific notation is used to make it easier to express extremely large or extremely small numbers. This notation is based on multiplying a number by a power of ten ($\times 10^{n}$).
- Expressing numbers in scientific notation makes it easier to perform simple mathematical operations on that number.

Worked example

The oceans of the world contain both very large and very small organisms. Orders of magnitude can be used to compare organisms that vary hugely in size. Take, for example, the masses of the following organisms:

- blue whale = 140 000 kg
- single plankton = 0.5 mg

Expert tip

'Weight' and 'mass' do not mean the same thing. The mass of an object is the amount of matter an object is made up of, which does not change. The weight of an object is due to the gravitational force acting on the object, and so can vary depending on where the object is in the universe and the gravity acting on it. The mass of an organism is measured in kg whereas its weight/weight force is measured in newton (N). On the surface of the Earth an object with a mass of 1 kg has a weight of about 10 N.

To calculate the order of magnitude difference between these two organisms, the masses need to be expressed in the same unit (kilogram):

- blue whale = 140 000 kg = 1.4×10^5 kg
- plankton = 0.0000005 kg = 5×10^{-7} kg

To compare directly the weights of these two organisms, use the following calculation:

$$\frac{1.4 \times 10^5}{5 \times 10^{-7}} = 0.28 \times (10^5 \times 10^7)$$

$$= 0.28 \times 10^{12} = 2.8 \times 10^{11}$$

This means that the blue whale is approximately 10^{11} times more massive than the plankton (× 100 000 000 000), that is, 11 orders of magnitude heavier.

Figure 1.2 shows the variation in scale in biology by comparing the size of different biological objects to one of the smallest (a rhinovirus).

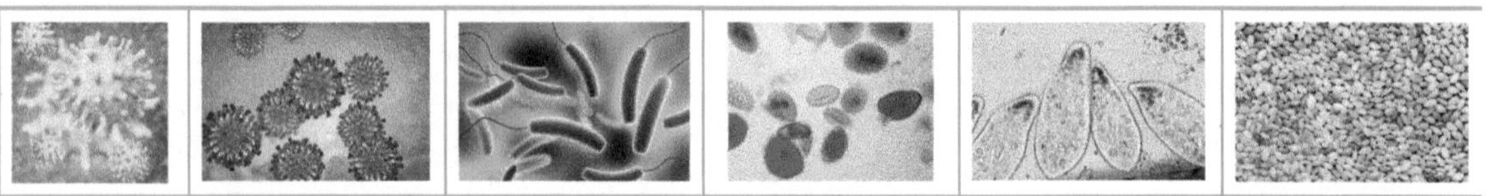

Name	Rhinovirus	Measles virus	*E. coli*	Pollen grain	*Paramecium*	Sesame seed
Length	20 nm	200 nm	2 µm	20 µm	200 µm	2 mm
	2×10^{-8} m	2×10^{-7} m	2×10^{-6} m	2×10^{-5} m	2×10^{-4} m	2×10^{-3} m
Order of magnitude	0	1	2	3	4	5

Woodlouse	Meerkat	Human	Sperm whale	Redwood tree	Seagrass colony	Great Barrier Reef
20 mm	200 mm	2 m	20 m	100 m	8 km	2 000 km
2×10^{-2} m	2×10^{-1} m	2×10^0 m	2×10^1 m	1×10^2 m	8×10^3 m	2×10^6 m
6	7	8	9	10	11	14

Figure 1.2 Sizes of different biological objects and differences in order of magnitude (bottom row) compared with a rhinovirus

■ **ACTIVITIES**

6   How many orders of magnitude larger is 5.42×10^7 compared with 4.70×10^{-3}?

7   How many times smaller is a ribosome (20 nm) compared with the length of an *Escherichia coli* bacterium (2 μm)? How many orders of magnitude smaller is the ribosome?

The following videos and sites explore order of magnitude:

- https://www.youtube.com/watch?v=0fKBhvDjuy0
- https://www.youtube.com/watch?v=jfSNxVqprvM
- http://learn.genetics.utah.edu/content/cells/scale/
- https://www.cellsalive.com/howbig_js.htm

Concept of significant figures

The number of significant figures (sf) in a numerical result is an indication of the accepted error in a measurement. The result of a calculation that involves measured values cannot be more certain than the least certain of the data that are used. Therefore, the result should contain the same number of significant figures as the measurement that has the smallest number of significant figures.

The following rules should be applied to establish the number of sf in a number:

■ Zeros between digits are significant. For example, 2 006 g has four significant figures.

■ Zeros to the left of the first non-zero digit are not significant (even when there is a decimal point in the number). For example, 0.005 g has one significant figure.

■ When a number with a decimal point ends in zeros to the right of the decimal point, these zeros are significant. For example, 2.0050 g has five significant figures.

■ When a number with no decimal point ends in several zeros, these zeros might or might not be significant. The number of significant figures should then be stated. For example, 30 000 g (to 3 sf) means that the mass has been measured to the nearest 100, while 30 000 g (to 4 sf) means that the mass has been measured to the nearest 10.

When significant figures are used as an implicit way of indicating uncertainty (see Chapter 10, pages 129–31), the last digit is considered uncertain. For example, a result reported as 1.23 implies a minimum uncertainty of +0.01 (or a maximum uncertainty of ±0.05) and a range of 1.22 to 1.24. Figure 1.3 shows the concept of uncertainty, which is explored in more detail later in this book (page 129).

Figure 1.3 A magnified thermometer scale showing a temperature of 18.7 °C: the last digit is uncertain

■ **ACTIVITY**

8   **State** and **explain** the number of significant figures in the following measurements:

14.44, 9 000, 3 000.0, 1.046, 0.26 and 6.02×10^{23}.

Examiner guidance

For multiple calculations, compute the number of significant digits to retain in the same order as the operations: first logarithms and exponents, then multiplication and division, and finally addition and subtraction. When parentheses are used, do the operations inside the parentheses first.

Expert tip

Place holder zeros can be removed by converting numbers to scientific notation. For example, 2 000 may have anywhere from one to four sf, but by writing the number in scientific notation the number of sf is made explicit; for example, by writing the number as 2.00×10^3 it is made clear that it has 3 sf.

Expert tip

30–300 rule

This rule is used to determine how accurately to measure a variable. The number of significant digits should be such that there are 30 to 300 units (approximately) between the largest and smallest measurement. For example: when measuring sardine lengths that range between 4 and 8 cm, there are only 4 cm between the largest and smallest values. The degree of accuracy that 1 cm intervals provide is not adequate. If the sardines are measured to the nearest 0.1 cm between 4.0 and 8.0 cm, there are 40 units of 0.1 cm between the largest and smallest values.

■ Rounding off significant figures

Sometimes it is necessary to round off, to give the correct number of significant figures.

- ■ A digit of 5 or larger rounds up.

- ■ A digit smaller than 5 rounds down.

The number 350.99 rounded to:

4 sf is 351.0

3 sf is 351

2 sf is 350

1 sf is 400

Notice that when rounding you only look at the one figure beyond the number of figures to which you are rounding, that is, to round to three significant figures you only look at the fourth figure.

Rounding depends on the number of significant figures allowed by the accuracy of the initial measurements.

> **Common mistake**
>
> A common mistake is to simply copy down the final answer from the display of a calculator. This often has far more significant figures than the measurements justify and you will lose marks for this under the Analysis and Communication criteria of your IA.

■ ACTIVITY

9   Report the following numbers to three significant figures:

654.389

65.4389

654,389

56.7688

0.03542210

Measurement: area, volume, mass, density and temperature

■ Area

The metre (m) is the base unit of length. Units of area are squared units of length (m^2), and so area is a two-dimensional measurement.

Surface area is an important concept in biology.

- ■ Total skin area of adult human = 1.8 m^2

- ■ Surface area of human lungs = 80 m^2

■ Volume

Volume is the space occupied by an object. The volume of a cube is given by its length cubed, that is, length^3, and so is a three-dimensional measurement. This means that the base SI unit of volume is the cubic metre (m^3) – that is, the volume of a cube that is 1 m on each edge. Smaller units, such as cubic centimetres, cm^3 (sometimes written as cc, although this should be discouraged), are often used in biology. A litre (L) is equivalent to a cubic decimetre, dm^3. As discussed above, the litre is not a standard SI unit, but is frequently used in biology.

- ■ 1 000 millilitres (ml) = 1 litre

- ■ Each millilitre is the same volume as a cubic centimetre: 1 ml = 1 cm^3.

Various devices are available to move and deliver, or measure, a volume of liquid:

■ A pipette can be used to extract and deliver volumes of liquid.

　□ They are used to move small volumes, typically 25 ml or less.

　□ A suction bulb draws fluid into the pipette.

　□ The mouth should never be used to suck fluid into a pipette.

　□ Graduated markings on a graduated pipette allow precise measurement of the volume of a liquid.

■ A graduated cylinder or measuring cylinder can be used to measure volumes.

　□ They are used to measure larger volumes, or when the precision of the measurement is less critical.

　□ The measurement should always be taken from the bottom of the meniscus (the interface between the water and air – it is curved because of surface tension and the adhesion of water to the sides of the cylinder) – see Figure 1.4. **Parallax error** occurs when there is a displacement or difference in the apparent position of an object viewed along two different lines of sight.

　□ Care should be taken to remove as much of the liquid from the cylinder as possible, although there will always be some remaining (a limitation of this technique).

■ Burettes can also be used to measure volumes of a liquid.

　□ They are similar to a graduated cylinder but have a stopcock at the bottom.

　□ They can be used to transfer liquids but are mainly used in titrations.

　□ The flow of liquid is controlled using the stopcock – it can be left completely open to allow a continuous flow or set to release one drop at a time.

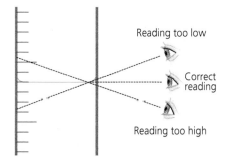

Figure 1.4 Parallax error with a measuring cylinder

> **Expert tip**
>
> When measuring the liquid in a cylinder such as a graduated cylinder, the meniscus should be read with eyes level with the meniscus. Read the volume at the lowest level.

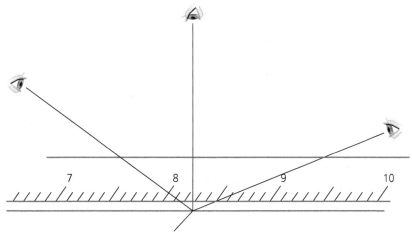

Figure 1.5 Parallax error when reading a metre ruler

> **Expert tip**
>
> Parallax error must also be considered when using a ruler (see Figure 1.5).

Measuring small volumes

Many biochemical measurements are made on a very small scale using automated micropipettes (Figure 1.6), many of which are air displacement piston pipettes. They are quick and convenient, but their accuracy is dependent on how they are adjusted and used. They come in a range of sizes, but can measure volumes on the microlitre scale.

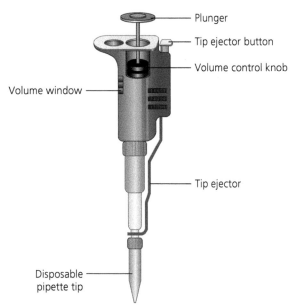

Plunger

Tip ejector button

Volume control knob

Volume window

Tip ejector

Disposable pipette tip

Figure 1.6 Automated micropipette

Magnetic stirrers

Magnetic stirrers are used for the mixing of large volumes (Figure 1.7) of solutions. Magnetic stirrers contain an electric motor that rotates a strong permanent magnet. The spin speed of the motor can be adjusted. The glassware containing the liquid to be mixed is placed on top of the stirrer. Some stirrers are equipped with a heater (hot plate) to warm up the solution. A stirring bar (a strong bar magnet with plastic coating) is dropped into the liquid.

Accurate measurement of liquids

High viscosity liquids, such as concentrated sucrose solution, are difficult to transfer; allow time for all the liquid to transfer.

Organic solvents or hot liquids may evaporate quickly, making measurements inaccurate; transfer these liquids quickly and cover containers with lids or parafilm.

Liquids likely to froth, for example, yeast or protein solutions, are difficult to measure; transfer slowly.

Suspensions, for example, yeast or cell cultures, may sediment; mix well before transferring.

Use measuring cylinders on a level surface so the scale is horizontal and parallax error is avoided (see above); fill to below the desired mark, then add liquid or solution slowly, for example, by pipette, to reach the desired level.

Make sure there are no air bubbles in syringes when measuring volumes. Expel liquids slowly and touch the end of the syringe on the vessel to remove any liquid or solution stuck to the end.

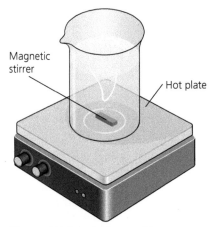

Magnetic stirrer

Hot plate

Figure 1.7 A hot plate with magnetic stirrer

■ ACTIVITY

10 Identify or correct the mistakes in the following: 5.0 secs, 51.2 cms, 20.0 × 40.0 cm, speed is 0.01 ms⁻¹, weight of the beetle is 4.25 g.

Mass

The kilogram (kg) is the basic unit of mass. Electronic balances are usually used to record the mass of an object.

■ Before making any measurements, clean the weighing pan.

■ Zero the balance, so it indicates '0 g' (zero grams).

■ Measure the mass of an object by placing it in the centre of the weighing pan.

■ Take care to ensure the balance is zeroed between each measurement.

■ Some balances have doors that need to be shut, to ensure air movement does not affect measurements.

Density

Density is mass per unit volume (that is, mass/volume). The average density of an object equals its total mass divided by its total volume. The SI unit for density is $kg\,m^{-3}$.

If solid objects sink when placed in water, they have a density greater than water ($1\,000\,kg\,m^{-3} \equiv 1\,g\,ml^{-1}$).

If several liquids that do not mix (that is, they are immiscible) are placed in the same vessel, the densest one will sink to the bottom and the least dense will rise to the top.

Temperature

Temperature is a measure of the average kinetic energy of molecules in a system. Biologists measure temperature using a thermometer calibrated in degrees Celsius (°C). The Celsius scale is based on water freezing at 0 °C and boiling at 100 °C at 1 atm pressure. Digital thermometers are the most accurate, and safest, way of recording temperature (Figure 1.8).

Figure 1.8 A digital thermometer being used to record soil temperature

Expert tip

'Heat' and 'temperature' are often used interchangeably, but they are actually different, although closely related, concepts. Temperature is a measure of the average energy of molecular motion in a substance whereas heat is the total energy of molecular motion in a substance. Temperature is measured in degrees Celsius; heat is measured in joules.

Expert tip

Degrees Celsius (°C) and degrees Fahrenheit (°F) can be interconverted using the formula:

$$°F = \frac{9}{5}\,°C + 32$$

2 Biochemistry practical skills

Note: Before you embark on a practical biochemistry investigation, you must carry out a full risk assessment and gain approval from your teacher (see pages 37–9). Consult the manufacturer's operating instructions for instruments and equipment or published papers with protocols.

Food tests

Chemical tests can be carried out that identify biological molecules such as starch, disaccharides and monosaccharides. Lab-made solutions of the chemical compounds can be tested as well as samples from actual food. Reagents can be used to test for different food groups.

- Starch can be tested for using iodine: the polysaccharide forms a blue-black starch–polyiodide complex.

- Monosaccharides and disaccharides can be tested for using the Benedict's test. Benedict's reagent is an aqueous solution of copper(II) sulfate (copper(II) ions), sodium carbonate and sodium citrate. All monosaccharides and most disaccharides (except sucrose) will reduce copper(II) sulfate, producing an orange/red precipitate of copper(I) oxide on heating, so they are called reducing sugars. The colour and density of the precipitate give an indication of the amount of reducing sugar present, so this is a semi-quantitative test.

- Because sucrose does not reduce copper(II) sulfate (copper(II) ions) on heating, it is called a non-reducing sugar. If a non-reducing sugar is broken down into its constituent monosaccharide sugar molecules (glucose and fructose), via hydrolysis, then these sugars can be tested for using the Benedict's test, giving a positive test.

Equipment

- test tubes
- test tube rack
- dropping pipette
- lab-made solutions: starch, glucose (or another monosaccharide) in 0.1 %, 1 % and 10 % solution, and sucrose
- Bunsen burner (or water bath)
- tripod and gauze

- heatproof mat
- test tube holder
- eye protection
- iodine solution
- Benedict's reagent
- distilled water
- dilute hydrochloric acid.

Safety

- Ensure eye protection is worn.
- Take care using hydrochloric acid.
- Procedure uses hot water bath – take care when adding and removing tubes.

Testing for starch

1 Add two drops of iodine solution to 2 cm³ of test solution.

2 A blue-black colour indicates the presence of starch.

3 Starch is only slightly soluble in water but the test works well in a suspension or as a solid.

Testing for reducing sugars

1 Add 5 cm³ of Benedict's reagent to 5 cm³ of the solution to be tested
   (0.1 % glucose, 1 % glucose, 10 % glucose and 10 % sucrose).

2 Add 5 cm³ of Benedict's reagent to 5 cm³ of distilled water (control experiment).

3 Put the five test tubes in a water bath at about 95 °C for five minutes.

4 Transfer the test tubes to a rack and compare the colours. The original pale
   blue colour means no reducing sugar; a green precipitate means relatively
   little sugar; a brown or red precipitate means progressively more sugar is
   present.

5 0.1 % glucose should show little colour change; 1.0 % glucose should show
   some colour change (green/orange); and 10 % glucose should show the largest
   colour change (orange/red). There should be no colour change with sucrose or
   distilled water.

6 Keep the test tubes so that they can be compared with the test for non-
   reducing sugars.

Testing for non-reducing sugars

1 Boil 5 cm³ of the test solution with 2 cm³ dilute hydrochloric acid (a source of
   hydrogen ions, H⁺(aq)) for a few minutes: this process hydrolyzes the glycosidic
   bond.

2 Slowly add small amounts of solid sodium hydrogencarbonate to the solution,
   until it stops fizzing: this neutralizes the solution.

3 Carry out the test for reducing sugars.

4 Compare the amount of precipitate formed with the results from the reducing
   sugar tests – the quantity of precipitate produced indicates the amount of
   monosaccharides present (and so, indirectly, the quantity of non-reducing
   sugar in the solution).

Testing food samples

In these tests, a piece of food is taken and ground with a pestle and mortar to
break up the cells and release the cell contents. Many of these compounds are
insoluble, but the tests work just as well on a fine suspension.

Equipment

- test tubes
- test tube rack
- glass rod
- dropping pipette
- spatula
- pestle and mortar
- small beaker
- Bunsen burner (or water bath)

- tripod and gauze
- heatproof mat
- test tube holder
- eye protection
- iodine solution
- Benedict's reagent
- distilled water
- dilute hydrochloric acid

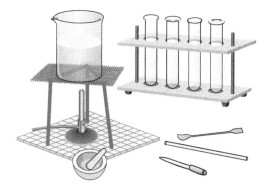

Figure 2.1 Equipment needed to test foods
for chemical compounds

Safety

■ Ensure eye protection is worn.

■ Take care using hydrochloric acid.

■ Procedure uses hot water bath – take care when adding and removing tubes.

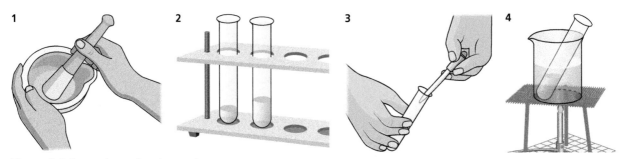

Figure 2.2 Preparing a food sample to test

1 Make an extract by removing a small amount of the food sample and then grinding it using a pestle and mortar to produce (if possible) a fine powder.

2 Put the ground-up food sample into a test tube to a depth of about 2 cm. Add a similar amount of distilled water and stir with a glass rod. Allow to stand for a few minutes.

3 Draw up about 2 cm³ of the clear liquid into a pipette. If testing for reducing sugars, add the food solution to a test tube containing 2 cm³ of Benedict's solution. Carry out steps 4 and 5.

4 Put the test tube into a beaker of boiling water and leave for a few minutes so that the mixture is thoroughly heated.

5 Remove the tube using a test tube holder. A change in colour from blue to yellow-red indicates the presence of reducing sugar (see Figure 2.3 below).

6 If testing for non-reducing sugars, test a sample for reducing sugars, to see if there are any present. Boil the test solution with dilute hydrochloric acid for a few minutes to hydrolyze the disaccharide sugar. Neutralize the solution by carefully adding small amounts of solid sodium hydrogencarbonate until a neutral pH is achieved. Test as before for reducing sugars.

7 If testing for starch, take 2 cm³ of the food solution and add a few drops of iodine solution – a change to blue-black colour indicates the presence of starch.

Ideas for investigations

The sugar content of foods such as fruit is affected by the conditions in which it is kept, and the age of the fruit. The concentrations of reducing sugars, under these different conditions, can be determined semi-quantitatively using Benedict's reagent and a range of standard glucose solutions.

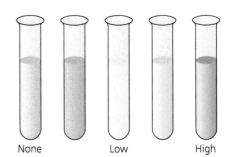

None   Low   High

Figure 2.3 Colour changes using Benedict's test indicate the approximate quantity of monosaccharides present

Practical 3: Experimental investigation of a factor affecting enzyme activity

The aim of these experiments is to:

- Carry out an investigation into how one factor affects **enzyme** activity. Factors can include temperature, pH or **substrate** concentration.

- Make sure that accurate, quantitative measurements are taken by using replicates to ensure reliability.

Experiment 1: The effects of temperature on hydrolysis of starch by amylase

When starch is hydrolyzed by water in the presence of the enzyme amylase, the product is maltose, a disaccharide. Starch gives a blue-black colour when mixed with iodine solution (iodine in potassium iodide solution) but maltose gives a red colour.

Equipment

- amylase 0.1 %
- starch 1 %
- iodine
- white spotting tiles
- dropper
- stop clock
- test tubes

Procedure

Carry out the experiment as shown in Figure 2.4.

1   Bring samples of the enzyme and the substrate (the starch solution) to the temperature of the water bath before mixing them: this is called pre-incubation.

2   Follow the progress of the hydrolysis reaction by taking samples at half-minute intervals. For each sample, take a drop of the mixture on the end of the glass rod and transfer it to a white spotting tile. Test each sample with iodine solution.

3   Initially, a strong blue-black colour is seen, confirming the presence of starch. Later, as maltose accumulates, a red colour predominates.

4   The endpoint of the reaction is when all the starch colour has disappeared from the test spot.

5   Using a fresh reaction mixture each time, repeat the investigation at a series of different temperatures, for example, at 10 °C, 20 °C, 30 °C, 40 °C, 50 °C and 60 °C. Record the time taken for complete hydrolysis at each temperature and plot the rate of hydrolysis per unit time on a graph.

> **Key definitions**
>
> **Enzyme** – usually a protein (a very few are RNA) that functions as a biological **catalyst**.
>
> **Substrate** – a molecule that is the starting reactant for a biochemical reaction and that forms a complex with the active site of a specific enzyme.

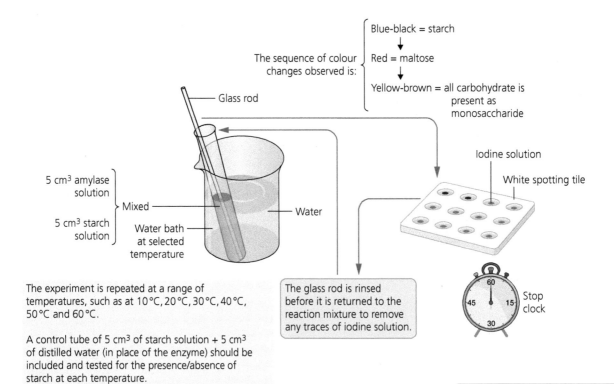

The sequence of colour changes observed is:

Blue-black = starch
↓
Red = maltose
↓
Yellow-brown = all carbohydrate is present as monosaccharide

Glass rod

Iodine solution

White spotting tile

5 cm³ amylase solution

5 cm³ starch solution

Mixed

Water

Water bath at selected temperature

The experiment is repeated at a range of temperatures, such as at 10 °C, 20 °C, 30 °C, 40 °C, 50 °C and 60 °C.

A control tube of 5 cm³ of starch solution + 5 cm³ of distilled water (in place of the enzyme) should be included and tested for the presence/absence of starch at each temperature.

The glass rod is rinsed before it is returned to the reaction mixture to remove any traces of iodine solution.

Stop clock

Figure 2.4 The effects of temperature on the hydrolysis of starch by amylase

Results

Examine the results of the effect of temperature on the hydrolysis of starch by the enzyme amylase, shown in Figure 2.5 below.

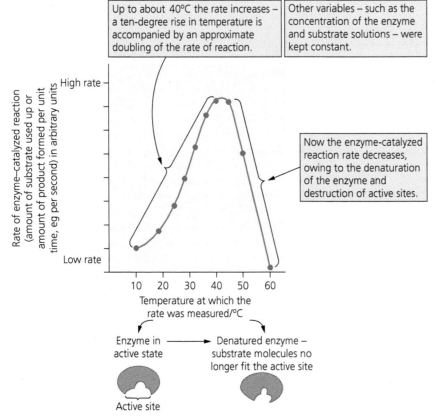

Up to about 40°C the rate increases – a ten-degree rise in temperature is accompanied by an approximate doubling of the rate of reaction.

Other variables – such as the concentration of the enzyme and substrate solutions – were kept constant.

High rate

Rate of enzyme-catalyzed reaction (amount of substrate used up or amount of product formed per unit time, eg per second) in arbitrary units

Now the enzyme-catalyzed reaction rate decreases, owing to the denaturation of the enzyme and destruction of active sites.

Low rate

10   20   30   40   50   60
Temperature at which the rate was measured/°C

Enzyme in active state → Denatured enzyme – substrate molecules no longer fit the active site

Active site

Figure 2.5 Temperature and the rate of an enzyme-catalyzed reaction

The results show a characteristic curve, although the optimum temperature varies from reaction to reaction and with different enzymes.

Analysis

- As temperature is increased, molecules have increased kinetic energy, and reactions between them go faster.

- The molecules are moving more rapidly and are more likely to collide and react. Q_{10}, or temperature coefficient, is a measure of the rate of change of a reaction when the temperature is increased by 10 °C. Many enzymes have a Q_{10} of about 2, which means that in many chemical reactions the rate of the reaction approximately doubles for every 10 °C rise in temperature.

- However, in enzyme-catalyzed reactions the effect of temperature is more complex, as proteins are denatured by heat. The rate of **denaturation** increases at higher temperatures. So, as the temperature rises the amount of active enzyme progressively decreases, and the rate is slowed. As a result of these two effects of temperature on enzyme-catalyzed reactions, there is an apparent optimum temperature for an enzyme. Humans have enzymes with optima at or about normal body temperature.

> **Key definition**
>
> **Denaturation** – a conformational change in a protein that results in a loss (usually permanent) of its biological properties.

Worked example

The following results were obtained in an investigation of the effects of pre-incubation of starch and amylase solutions at different temperatures on the subsequent hydrolysis of the starch to maltose.

Temperature of solution/°C ± 0.1 °C	10.0	20.0	30.0	35.0	40.0	45.0	50.0	55.0	60.0
Time/s ± 1 s	100	58	30	21	15	11	19	46	100

If you were asked to plot a graph of these results and analyse the trend shown, you would draw a graph like the one shown in Figure 2.6.

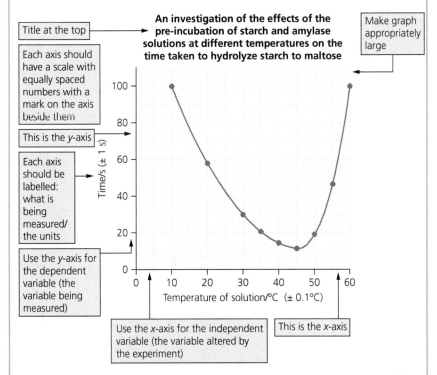

Figure 2.6 Graph showing the relationship between temperature and the time taken to hydrolyze starch to maltose

Expert tip

If the reciprocal of time was plotted on the *y*-axis an inverted U-shape would be obtained.

Between 10.0 and 45.0 °C, the time taken to hydrolyze the starch decreased. This is because the kinetic energy of the molecules increased, leading to increased collisions between the enzyme and substrate (per unit time), and more enzyme–substrate complexes formed. The graph shows that the optimum temperature for the enzyme is 45.0 °C. Between 50.0 and 60.0 °C, the time taken to hydrolyze the starch increased, as the amylase started to denature and fewer enzyme–substrate complexes formed (per unit time).

Experiment 2: The effects of pH on enzyme activity

Changes in pH can have a dramatic effect on the rate of an enzyme-catalyzed reaction:

- Each enzyme has a range of pH in which it functions efficiently.

- pH affects enzymes because the structure of a protein (and, therefore, the shape of the **active site**) is maintained by various bonds within its three-dimensional structure.

- A change in pH from the optimum alters the bonding patterns, progressively changing the shape of the molecule:

  - Acidity is due to the presence of hydrogen ions (H^+), and alkalinity can be due to the presence of hydroxide ions (OH^-). H^+ and OH^- ions are charged and therefore interfere with the hydrogen and ionic bonds that hold an enzyme together and determine its tertiary structure, since the ions will be attracted or repelled by the charges created by the bonds.

  - The active site may quickly be rendered inactive and unable to bind to the substrate.

  - Unlike temperature changes, the effects of pH on the active site are normally reversible – provided the change in surrounding acidity or alkalinity is not too extreme; as the pH reverts to the optimum for the enzyme, the active site may reform.

> **Key definition**
>
> **Active site** – region of enzyme molecule where the substrate molecule binds and catalysis occurs.

Some of the digestive enzymes of the gut have different optimum pH values from the majority of other enzymes. For example, those adapted to operate in the stomach, where there is a high concentration of hydrochloric acid during digestion, have an optimum pH which is close to pH 2.0 (Figure 2.8).

Equipment

- amylase 1 % (or 0.5 %)

- starch 1 % (or 0.5 %)

- buffer solutions covering a range of pH (for example, 2–9), each with a labelled syringe/plastic pipette

- two 5 cm³ syringes (one for starch, one for amylase)

- iodine solution in a dropper bottle

- test tubes (one for each pH to be tested) and a test tube rack

- spotting tile

- teat pipette

- stop clock

- marker pen

Procedure

1 Place a single drop of iodine solution in each dimple of the spotting tile.

2 Label a test tube with the pH to be tested.

3 Use a syringe to place 2 cm³ of amylase into the test tube.

4 Using a different syringe, add 1 cm³ of buffer solution to the test tube.

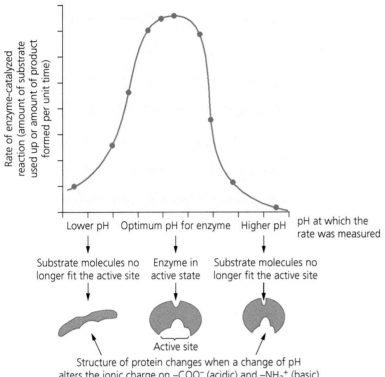

Figure 2.7 Effect of pH on enzyme shape and activity

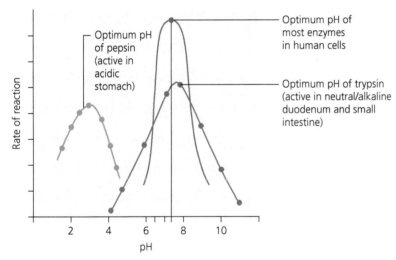

Figure 2.8 The optimum pH of different human enzymes

5 Use another syringe to add 2 cm³ of starch to the amylase/buffer solution. Start the stop clock.

6 After 10 seconds, use the plastic pipette to place one drop of the mixture into the first dimple containing iodine. The iodine solution should turn blue-black. Put the remaining solution back into the amylase/buffer/starch solution, leaving the pipette empty.

7 Wait another 10 seconds, and then remove a second drop of the mixture to add to the next dimple containing iodine.

8 Repeat until the iodine solution and the amylase/buffer/starch mixture remain orange. (*You could prepare a control drop for comparison with the test drops. What should this contain?*)

9 Count how many iodine dimples you have used up to the point at which the iodine remained orange: each one equals 10 seconds of reaction time.

10 Plot a graph of your results.

Analysis

- What is the optimum pH for the amylase? How do you know?

- What happened at pH values above and below this optimum?

- Use your knowledge of enzyme theory to fully explain your results.

Experiment 3: The effect of catalase concentration on the breakdown of hydrogen peroxide

The effect of different concentrations of substrate on the rate of an enzyme-catalyzed reaction can be shown using an enzyme called catalase. This enzyme catalyzes the breakdown of hydrogen peroxide:

$$2H_2O_2(aq) \rightarrow 2H_2O(l) + O_2(g)$$

The function of catalase is to protect the biochemical machinery of cells. Hydrogen peroxide is a common by-product of reactions of **metabolism**: it is a very powerful oxidizing agent and therefore potentially a very toxic substance within cells. Catalase inactivates hydrogen peroxide as it forms, before damage can occur.

The effect of substrate concentration on enzyme activity can be measured in one of two ways:

1   Measure the amount of substrate that has been converted from a reaction mixture in a specific amount of time.

2   Measure the amount of product that has accumulated in a unit of time.

The reaction between catalase and hydrogen peroxide produces oxygen, so in this experiment it is convenient to measure the rate at which oxygen accumulates. The volume of oxygen that has accumulated at 30-second intervals can be recorded as shown in Figure 2.9. The concentration of hydrogen peroxide used is 3 %, which is equivalent to 0.88 mol dm^{-3} by mass.

The graph in Figure 2.9 shows the following:

- Over a period of time, the initial rate of reaction is not maintained but falls off quite sharply. This is typical of enzyme actions studied outside their location in the cell.

- The decrease can be due to a number of reasons but most commonly it is because the concentration of the substrate in the reaction mixture has decreased.

Because the rate of reaction decreases, the initial rate of reaction is measured in such experiments: this is the slope of the tangent to the curve in the initial stage of the reaction.

Equipment

- measuring cylinder

- test tubes

- delivery tube

- hydrogen peroxide (3 % is the most widely available)

- catalase (if pure enzyme is not available, use a potato cut into small cubes)

- glass trough (to submerge the inverted measuring cylinder in – see Figure 2.9)

- stop clock

Safety

Hydrogen peroxide is a toxic substance (although in this experiment it is heavily diluted):

- it is a strong oxidizing agent

- contact with other material may cause a fire

- eye contact may result in permanent eye damage

- causes eye and skin irritation and possible burns

- corrosive

- may cause severe respiratory tract irritation with possible burns

- may cause severe digestive tract irritation with possible burns.

Procedure

To investigate the effects of substrate concentration on the rate of an enzyme-catalyzed reaction, the experiment shown in Figure 2.9 is repeated at different concentrations of substrate, and the initial rate of reaction is plotted in each case. Other variables such as temperature and enzyme concentration are kept constant.

The test tube is tipped up to mix the enzyme solution with the substrate.

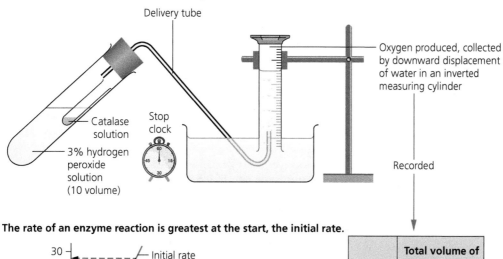

The rate of an enzyme reaction is greatest at the start, the initial rate.

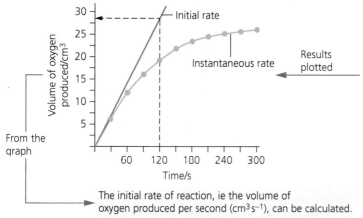

Time/s (± 1 s)	Total volume of oxygen gas collected/cm³ (± 0.5 cm³)
30	6.0
60	12.0
90	16.0
120	19.0
150	22.0
180	23.0
210	24.0
240	25.0
270	25.5
300	26.0

The initial rate of reaction, ie the volume of oxygen produced per second (cm³ s⁻¹), can be calculated.

Figure 2.9 Measuring the rate of reaction using catalase

1 Prepare six concentrations of hydrogen peroxide as indicated in Figure 2.10.

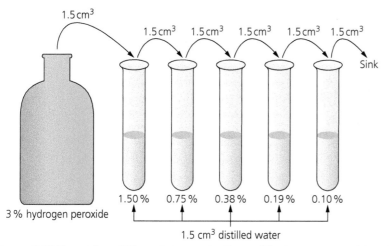

Figure 2.10 Preparing different concentrations of hydrogen peroxide using serial dilution

2 Add hydrogen peroxide to the apparatus shown in Figure 2.9 (start with either the highest or lowest concentration).

3 Mix the catalase with the hydrogen peroxide and measure the volume of oxygen produced in a fixed amount of time.

4 Record the results and repeat the experiment with fresh glassware, using the next hydrogen peroxide concentration.

5 Ensure that you repeat five times for each substrate concentration.

Experimental method

Accurate measurements are needed in scientific experiments. Accuracy relates to how close your results are to the true value. Accuracy can be improved by carefully measuring time and volume in this experiment and averaging precise volumes. Limitations in your equipment and method will reduce the accuracy of the results. For example, solutions of hydrogen peroxide must be carefully prepared – if too much or too little water is mixed with the hydrogen peroxide then the concentration will be incorrect. Using a burette, rather than a pipette, to transfer the distilled water into the substrate solution will improve accuracy, because the water molecules will not stick to the sides of a burette as they might in a pipette.

Record your results in a table. The example below shows you how to arrange your table:

Hydrogen peroxide concentration (%)	Volume of oxygen produced per unit time/cm^3 (± 0.5 cm^3)					
	Repeat 1	Repeat 2	Repeat 3	Repeat 4	Repeat 5	Mean
0.10						
0.19						
0.38						
0.75						
1.50						
3.00						

Table 2.1 A table to use when recording results of an experiment investigating the effects of hydrogen peroxide concentration on volume of oxygen produced per unit time

Calculate the rate of reaction for each substrate concentration using the technique shown in Figure 2.9. Plot a graph of your results (see Figure 2.11).

Analysis

When the initial rates of reaction are plotted against the substrate concentration, the hyperbolic curve shows two phases:

▪ At lower concentrations, the rate increases in direct proportion to the substrate concentration.

▪ At higher substrate concentrations, the increase in the rate of reaction slowly decreases until it shows virtually no increase.

▪ An increasing substrate concentration causes a diminishing increase in the rate of reaction.

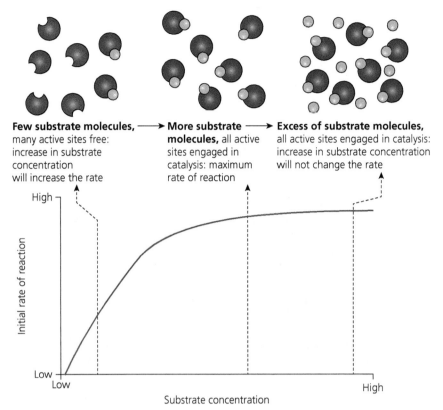

Few substrate molecules, ⟶ More substrate ⟶ Excess of substrate molecules,
many active sites free: **molecules,** all active all active sites engaged in catalysis:
increase in substrate sites engaged in increase in substrate concentration
concentration catalysis: maximum will not change the rate
will increase the rate rate of reaction

Figure 2.11 The effect of substrate concentration

Catalase works by forming a short-lived enzyme–substrate complex. At low concentration of substrate, all molecules can quickly diffuse and find an active site. Effectively, there is an excess of enzyme present. Here the rate of reaction is set by how much substrate is present – as more substrate is made available, the rate of reaction increases.

At higher substrate concentrations, there comes a point when there is more substrate than enzyme. Now, in effect, substrate molecules have to wait for access to an active site. Adding more substrate merely increases the number of molecules awaiting contact with an enzyme molecule, so there is no increase in the rate of reaction (Figure 2.11). This is known as saturation kinetics.

■ ACTIVITIES

1 **Explain** why the shape of globular proteins that are enzymes is important in enzyme action.

2 When there is an excess of substrate present in an enzyme-catalyzed reaction, **explain** the effect on the rate of reaction of increasing the concentration of: (a) the substrate; (b) the enzyme.

3 **Sketch** graphs to show the expected effects of temperature, pH and substrate concentration on the activity of enzymes. **Explain** the patterns or trends apparent in these graphs.

4 **Describe** and **explain** the process of denaturation.

5 Find out about the use of 'volume strength' as a measure of the concentration of hydrogen peroxide.

6 $0.50\ cm^3$ of blood is pipetted into $1.00\ cm^3$ of solution X. $0.025\ cm^3$ of a sample of solution X is pipetted into $1.00\ cm^3$ of solution Y. **Calculate** the total dilution factor for this assay.

Examiner guidance

A biochemical assay is a procedure for assessing or measuring the presence, amount or activity of a biochemical substance.

7 A control allows confirmation that no unknown or unidentified variable is responsible for any of the observed changes in the dependent variable. The use of a control ensures the experiment is a 'fair test'. **Outline** how a control could be used in an investigation of soybean digestion by the action of the enzyme papain.

Investigation into the effect of temperature on the breakdown of hydrogen peroxide

As we have already seen in the previous section, hydrogen peroxide (H_2O_2) is a toxic metabolic by-product that can cause damage to cells. The enzyme catalase is one of the most efficient enzymes known and catalyzes the breakdown of hydrogen peroxide into water and oxygen (page 22).

Potato cells can be used as a source of catalase enzyme. An easy way to vary the temperature at which the reaction takes place is to take equal-sized potato cubes and incubate them in thermostatically controlled water baths with hydrogen peroxide solution at different temperatures.

Equipment

- thermostatically controlled water baths at 20 °C, 30 °C, 40 °C, 50 °C, 60 °C and 70 °C
- pre-cut potato chips (rectangular prisms) with cross-sectional area 10 mm × 10 mm
- hydrogen peroxide stock solution (1.8 mol dm^{-3}/3 % concentration)
- boiling tube with delivery tube
- measuring cylinders for collecting oxygen gas
- stop clock
- forceps
- large beakers

Safety

- Take care cutting the chips (using either a chip-maker or knife). Cut away from the body onto a white tile or other cutting surface.
- Hydrogen peroxide is a toxic substance:
  - □ it is a strong oxidizing agent
  - □ contact with other material may cause a fire
  - □ eye contact may result in permanent eye damage
  - □ causes eye and skin irritation and possible burns
  - □ corrosive
  - □ may cause severe respiratory tract irritation with possible burns
  - □ may cause severe digestive tract irritation with possible burns.

Procedure

Set up the equipment as shown in Figure 2.12. In order to fill the measuring cylinder, fill it most of the way to the top with water. Place a finger over the top, invert the cylinder and place the open end under the water in the beaker. Remove your finger.

1   Add hydrogen peroxide to the reaction tube in a water bath and leave until the solution reaches the desired temperature (check this using a digital thermometer).

2   Add to the reaction tube a potato sample that has been pre-warmed ('incubated') to the same temperature as the hydrogen peroxide solution.

3   Record the starting level on the measuring cylinder.

4   Insert the bung into the reaction tube, ensuring you have a tight seal.

5   After 30 seconds place the end of the delivery tube under the inverted measuring cylinder. You will see bubbles of oxygen rising from the delivery tube.

6   Measure the volume of oxygen produced over a suitable time interval.

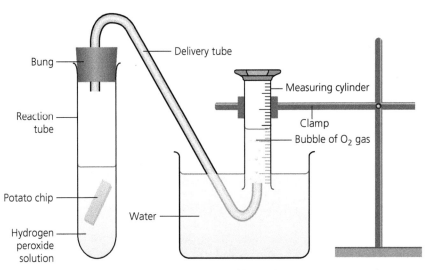

Figure 2.12 Experiment investigating the effect of temperature on the breakdown of hydrogen peroxide by catalase from a potato chip

Experimental method

Accurate measurements are needed in scientific experiments. Accuracy relates to how close your results are to the true value and whether they are within the limits of experimental error. Limitations in your equipment and method will reduce the accuracy of the results. For example, in this experiment the size and surface area of the chips will affect the rate of reaction – care must be taken to cut the chips to the same size. Another limitation that could affect the accuracy of the experiment is possible variation in catalase composition in different chips – the concentration of catalase cannot be assumed to be homogeneous in all potatoes. Systematic errors (see page 129) may also affect the results. For example, hydrogen peroxide decomposes spontaneously with time and is unstable with respect to decomposition. This could lead to different concentrations of the substrate being present in the experiment at different temperatures, rather than the assumed constant initial concentration (3 % solution/1.8 mol dm^{-3}). The precision of the experiment will depend on the smallest division marked on the graduated cylinder. An automated micropipette could be used (see page 12).

Results

Record your results in a table such as the one shown in Table 2.2.

Temperature/°C (± 0.1 °C)	Volume of oxygen produced in 1 minute/cm³ (± 0.5 cm³)					
	Repeat 1	Repeat 2	Repeat 3	Repeat 4	Repeat 5	Mean
20.0						
30.0						
40.0						
50.0						
60.0						
70.0						

Table 2.2 Example results table for an experiment investigating the effect of temperature on the decomposition of hydrogen peroxide

Expert tip

In an investigation, you should plot the variable you measured. You measured the volume of oxygen released in a set period of time, not the rate of reaction. You could calculate the rate of reaction using the method outlined on pages 22–3. When you manipulate data you should make it clear how you did this, and in a graph plotted from this manipulated data, you must make it clear that you have plotted processed data.

Analysis

In your analysis you need to discuss the following points:

- Why did the volume of oxygen increase as the temperature increased? (*Hint: talk about the kinetic energy of the colliding enzyme and substrate molecules, and the number of successful collisions per unit time.*)

- What was the optimum temperature of the enzyme? How do you know? Why did the volume of oxygen decrease above a certain temperature? (*Hint: write about what happens when an enzyme denatures.*)

■ ACTIVITY

8 In studies of the effect of temperature on enzyme-catalyzed reactions, **suggest** why the enzyme and substrate solutions are pre-incubated to a particular temperature before they are mixed.

Ideas for investigations

Enzymes provide many opportunities for carrying out biological investigations.

- Pectin is a substance that helps bind plant cells together. Pectinase breaks down pectin as the fruit ripens, causing it to soften. Pectinase can be bought for jam making. Investigations could explore the effect of pectinase on the fruit ripening process, or in juice extraction.

- Pineapples contain a protease enzyme, bromelain. This enzyme increases in concentration as the fruit ripens. An investigation could test pineapples of different ages for different concentrations of bromelain. Gelatin, which contains protein, could be used to test bromelain concentrations.

- Ascorbic acid oxidase is an enzyme released from fruits when their tissue is damaged. The enzyme is activated when exposed to air and causes tissues to lose vitamin C (ascorbic acid). Fruit juice is boiled soon after extraction to denature this enzyme and ensure vitamin C content is retained. An investigation could explore whether all fruits react in the same way, and how effective the process is.

Practical 4: Separation of photosynthetic pigments by chromatograph

Aims

To separate and identify the photosynthetic pigments present in at least three different plants.

To compare the photosynthetic pigments present in these plants.

Background information

Leaves contain many chemical compounds, such as pigments used in **photosynthesis**. For example, what is commonly (and inaccurately) known as 'chlorophyll' is actually a combination of five different pigments (or families of pigment): beta-carotene, chlorophyll *a*, chlorophyll *b*, pheophytins and xanthophylls.

Chromatography is used to separate pure substances from a mixture, such as a cell extract. This technique can be used to separate the photosynthetic pigments of a leaf so that they can be identified. Chromatography is based on different substances having different solubilities in different solvents. Solvents are usually mixtures of water with organic liquids such as ethanol or propanone. Extracts of plant material are dissolved in a solvent, which is then used to separate the different pigments through a medium such as filter paper. The relative distances travelled by different pigments depend on the type of solvent used. The simplest and most common form of chromatography uses filter paper, although other more sophisticated techniques are also used.

Expert tip

There are several different chromatography techniques that can be used to separate photosynthetic pigments. These include paper chromatography and thin layer chromatography. The procedure in this section covers paper chromatography, which is straightforward to carry out in a school or college laboratory. Other techniques are outlined in Figure 2.16 and at the end of the section.

Safety

- Follow the safety labels on each of the solutions or liquids.

- Solvent/propanone are both highly flammable and so must be handled with care.

- Do not light any matches near the solvents or use a Bunsen burner in the vicinity of the experiment.

- Keep in a sealed bottle when not in use and ensure that the room is well-ventilated.

- The solvent can cause severe eye damage, so eye protection must be worn.

- Propanone should be disposed of carefully and not washed down the sink.

- Centrifuge tubes must be balanced prior to centrifugation when using a centrifuge rotor that is symmetrical.

Procedure

1 Cut a strip of chromatography paper so that it fits into a gas jar.

2 Rule a pencil line across the strip of paper 30 mm from one end.

3 Pour some solvent into the gas jar and seal it, so the atmosphere is saturated with solvent vapour.

4 Using a mortar and pestle, grind up fresh leaves (for example, spinach) in propanone, producing as concentrated a pigment solution as possible (Figure 2.13). Filter the solution.

> **Expert tip**
>
> Whereas beta-carotene is an individual pigment, the xanthophylls include several different pigments, such as lutein, as do the pheophytins. The xanthophylls are oxidized versions of carotenes, and pheophytins are similar to chlorophyll but with the magnesium ion replaced by two hydrogen ions.

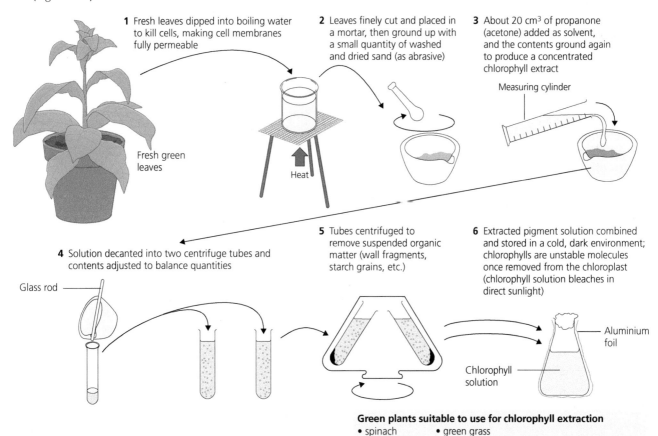

1 Fresh leaves dipped into boiling water to kill cells, making cell membranes fully permeable

2 Leaves finely cut and placed in a mortar, then ground up with a small quantity of washed and dried sand (as abrasive)

3 About 20 cm³ of propanone (acetone) added as solvent, and the contents ground again to produce a concentrated chlorophyll extract

Measuring cylinder

Fresh green leaves

Heat

4 Solution decanted into two centrifuge tubes and contents adjusted to balance quantities

5 Tubes centrifuged to remove suspended organic matter (wall fragments, starch grains, etc.)

6 Extracted pigment solution combined and stored in a cold, dark environment; chlorophylls are unstable molecules once removed from the chloroplast (chlorophyll solution bleaches in direct sunlight)

Glass rod

Aluminium foil

Chlorophyll solution

Green plants suitable to use for chlorophyll extraction
- spinach   • green grass
- bougainvillea   • in Asia, the leafy vegetables kai lan or kang kong
- hibiscus

Figure 2.13 Steps in the extraction of plant pigments

5 Using a very thin capillary tube or the head of a pin, place a drop of the pigment solution at the centre of the pencil line. This is the origin of the chromatogram.

6 Allow the pigment spots to dry (a fan can be used to speed up the process).

7 Place a second drop on the first. Repeat as many as 20 times so that you build up a small but concentrated spot of pigment.

8 Repeat steps 4 to 7 for any other plant samples. Label the spots with pencil (ink will dissolve in the solvent).

9 Using a ruler to measure distance, pour solvent into the gas jar to a depth of approximately 15 to 20 mm. Place a lid over the gas jar so that the atmosphere inside remains saturated with vapour.

10 Place the bottom edge of the chromatography paper into the solvent so that it is just touching. Make sure the pigment spots are not immersed: the chromatography sheet should be placed so that the origin is just above the level of solvent (see Figure 2.14). Leave for several hours.

11 The solvent front will rise up the paper by capillary action and separate the pigments. Any solutes dissolved in the solvent will be partitioned between the organic solvent (the moving phase) and the water, which is held by the surface of the paper (the stationary phase). The more soluble a solute is in the solvent, the further up the paper it will move.

12 When the solvent is close to the top of the paper, remove the strip and rule a pencil line to mark the distance travelled by the solvent.

13 The chromatogram might need to be developed to make the spots visible (for example, amino acids stain blueish-purple with ninhydrin).

14 Dry the paper.

15 Identify the pigments by their colours and their R_f values (see instructions below).

Sometimes chromatography using a single solvent does not separate all the constituents of a mixture. Two-dimensional chromatography can be used in these instances: the chromatography paper is turned through 90 degrees and run a second time using a different solvent. Solutes that were not separated by the first solvent should separate in the second because, once again, they have different solubilities.

Calculating R_f values

An R_f value is characteristic of a particular solute in a particular solvent at a particular temperature. It can be used to identify components of a mixture by comparing it with tables of known R_f values (see Table 2.3).

> **Expert tip**
>
> The retention factor (R_f value) is the ratio between the distance travelled by a solute and the distance travelled by the solvent. The ratio is independent of the length of a chromatogram and the distance travelled by the solvent front, and is therefore an accurate way of identifying substances.

The R_f value is calculated from the ratio of the distance from the origin to the centre of each spot divided by the distance travelled by the solvent front:

$$R_f = \frac{\text{distance pigment travels}}{\text{distance the solvent front travels}}$$

For example, in Figure 2.15 the R_f value would be 35.4 mm ÷ 79.0 mm = **0.45**

This indicates that the pigment is **chlorophyll *b*** (see Table 2.3).

If the solvent front is not a straight line (as in Figure 2.15), calculate the average distance travelled.

> **Expert tip**
>
> R_f values, like other ratios, do not have units and are expressed as a single number. R_f values vary between 0 and 1. A substance that is insoluble in the solvent has an $R_f = 0$ and will not move at all. A substance that is completely soluble has an $R_f = 1$ and will move the same distance as the solvent. R_f values will vary according to the solvent, or mixture of solvents, used.

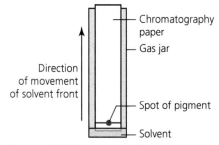

Figure 2.14 Creating a chromatogram

> **Expert tip**
>
> A suitable running solvent = petroleum ether + propanone in proportions of 2 : 1.

> **Expert tip**
>
> Prepare several chromatograms to ensure the reliability of your result. Take a photograph of the final results because the colours of the pigments fade quickly.

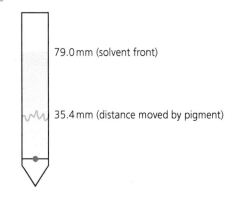

Figure 2.15 Calculating the R_f value for a pigment

Analysis and evaluation

◼ Examine your results and draw valid conclusions. Compare your results with values in the published literature (Table 2.3).

◼ Evaluate the procedure used and suggest improvements to the method.

Name	Colour	R_f value (2:1 petroleum ether + propanone)
Beta-carotene	Yellow	0.95
Pheophytin	Yellow-grey	0.83
Xanthophyll	Yellow-brown	0.71
Chlorophyll *a*	Blue-green	0.65
Chlorophyll *b*	Green	0.45

Table 2.3 R_f values for different pigments

Figure 2.16 shows a summary of the technique to measure R_f values using paper chromatography.

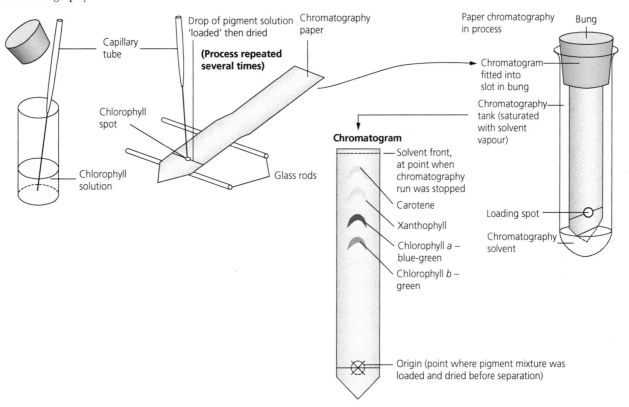

Figure 2.16 Preparing and running a chromatogram

Other chromatography techniques

Technique	Comment
Paper chromatography	Simplest technique, but does not always give very clean separation: it has lower sensitivity than TLC (see below) and therefore larger amounts of solute are needed.
Thin layer chromatography (TLC)	Uses a thin layer of cellulose, alumina or silica coated onto a plastic or glass sheet. More expensive, but more rapid and gives much better and more reliable separation (better resolution). There is less diffusion and more well-defined spots are formed.
High-performance liquid chromatography (HPLC)	Uses a technique called column chromatography, where a pump moves pressurized liquid solvent containing the sample mixture through a column filled with a solid adsorbent material. Delivers excellent separation very quickly.
Electrophoresis	Uses an electric current to separate molecules on the basis of charge. It can also be used to separate on the basis of molecular size. It is used in DNA sequencing.

Table 2.4 Different chromatography techniques

Expert tip

You will have heard the term 'absorbent' but might not be aware of the word 'adsorbent'. Adsorption and absorption are different physical and chemical processes. Absorption refers to the process by which a material soaks up an amount of liquid or gas into it, whereas adsorption is a process by which a liquid or gas accumulates on the surface of a solid material. Absorbent is a property related to volume and adsorbent to the surface material.

Expert tip

Paper chromatography can be used to separate photosynthetic pigments but thin layer chromatography gives better results because a thin adsorbent layer (for example, silica gel) runs faster and so has better separation.

Further biochemistry techniques and practices

This section includes commonly used techniques and practices, not included in experiments covered so far.

Centrifuges

Cells can readily be obtained from liquid cultures, and precipitated proteins in a solution can be separated into fractions by spinning the sample in a centrifuge (Figure 2.17). The fractions are referred to as the 'supernatant' (the solution) and the 'pellet' (the particles collected at the bottom of the centrifuge tube, pressed together into a small mass). The rotation generated by the electric motor of the centrifuge is transmitted to the rotor with the samples inside centrifuge tubes.

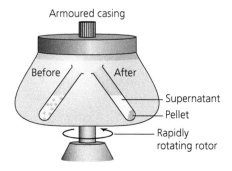

Figure 2.17 A centrifuge

Examiner guidance

The filled rotor must be counterbalanced: tubes with equal masses must be placed into opposite holes of the rotor. If the masses of the sample tubes are different, prepare counterbalances by filling tubes with water.

Homogenization

In this process, the cells are broken up completely so that a homogeneous mixture is obtained which can be oxygenated. The suspension medium can also be changed. The mincing of the material is done using either a homogenizer or a blender (Figure 2.18) which consists of a glass vessel fitted with rotating knives.

Storage of biological samples

Processing or purifying biological samples often takes time and manual work. It is often necessary to store the samples during an investigation. Ideally, the composition, biochemical and physical properties and the biological activity of the sample remain unchanged over time. It is very important to optimize storage conditions and minimize the time of storage.

Figure 2.18 A blender

A range of reactions, often under enzyme control, can occur in any biological sample at different rates, changing and/or degrading important chemicals. Some reactions require air, for example, oxygen reacts with sulfhydryl groups (–SH) of proteins. It is easy to protect the samples using an airtight cap. Further protection can be achieved by mixing additives directly into the sample. The rate of chemical reactions can also be decreased by lowering the temperature using a fridge.

Ultra-violet visible spectroscopy

Ultra-violet visible spectroscopy uses an instrument that measures the intensity of electromagnetic radiation which passes through an absorbing coloured aqueous solution. There is a linear relationship between the absorbance of a coloured solution and the concentration of the coloured substance (provided it is present at low concentration). Figure 2.19 shows the principle of a colorimeter. It can produce a beam of light of a given wavelength (using filters) and direct it at a sample in solution in a cuvette.

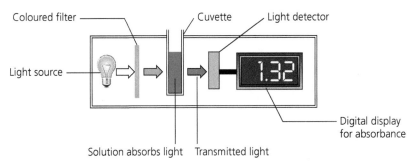

Figure 2.19 A colorimeter, showing the pathway of visible light (a blue or green filter is used with a red solution)

A spectrophotometer (ultra-violet radiation or visible light) or colorimeter (visible light) is an instrument for measuring the absorbance of a solution. Absorbance is (in a simple sense) a measure of the light absorbed by the sample that does not reach the detector.

Examiner guidance

Absorbance values greater than 2 are unreliable, because too little light is reaching the detector to allow for accurate measurements. When measuring absorbance, note the values; if the reading is greater than 2, dilute the sample by a factor of two, and repeat the measurement.

■ ACTIVITY

9 Colorimeters can be used to determine the approximate concentration of a highly coloured substance in solution. Find out how colorimeters can be used to estimate the concentration of food substances using the food tests (for example, iodine and Benedict's) outlined on pages 14–16.

■ Buffers

Proteins, and especially enzymes, are generally sensitive to changes in the concentrations of hydrogen ions and hydroxide ions. A buffer is a solution that is used to control the pH of a process occurring in an experimental aqueous medium.

Since most biochemical reactions occur in the aqueous medium of the cell, and since many reactions involve proton (H^+) transfers (acid–base reactions), pH is a very important variable in an enzyme-catalyzed reaction and must be controlled if pH is not an independent variable.

■ Gel electrophoresis

Expert tip

Electrophoretic separation occurs because of two principles: smaller fragments of DNA are less impeded by the gel matrix and so will move faster through the gel; the movement of fragments is possible because DNA is negatively charged due to the presence of a phosphate ion (PO_4^{3-}) on each nucleotide.

Gel electrophoresis (see Figure 2.20) is a technique used for separating fragments of DNA or proteins. The technique depends on the movement of charged molecules in an electric field. The polyacrylamide gel is used as the support medium and is contained in a tank. The relative rate of movement (for molecules of similar size) depends on the relative size of the DNA or protein molecule, with the smallest moving the furthest in a given time.

Expert tip

Always use a clean cuvette that has no scratches on it. Insert cuvettes correctly as they often have only two transparent sides for the light (or UV) to pass through.

Expert tip

When performing experiments with proteins, do not dilute the protein with water, unless denaturation of the protein is not an issue. When denaturation is likely, such as when performing dilutions for enzyme assays, perform the dilution using a suitable buffer to prevent unwanted changes in the shape of the protein in solution.

Key definition

Gel electrophoresis – a process that uses an electric field to separate proteins or fragments of DNA according to size.

Electrophoresis in progress

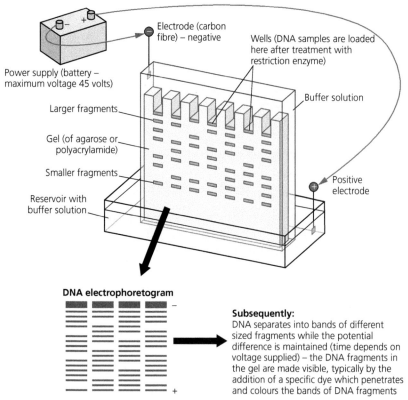

Figure 2.20 Gel electrophoresis

For protein separations, a buffer is employed, often coupled with a powerful detergent, for example, SDS (sodium dodecyl sulfate). A pattern of bands is revealed following staining (see Figure 2.21). For example, Coomassie Blue R-250, is commonly used to detect proteins. The proteins within an SDS polyacrylamide gel are denatured; the molecular mass determined will be that of the individual monomers of multimeric proteins.

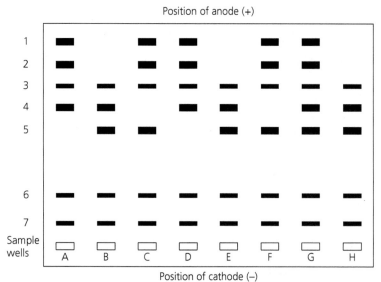

Figure 2.21 The results of an investigation into variation in liver enzymes in the house mouse, *Mus musculus*

For DNA separations, it is common to include in the buffer a denaturing agent (for example, urea) to keep the DNA in single-stranded form. The patterns of bands are compared with each other, or with known samples of proteins or DNA fragments.

◼ Purification

◼ Dialysis

Dialysis involves placing the protein solution in a semi-permeable membrane within a large container of buffer. Small particles (such as sodium and chloride ions) pass through the dialysis membrane (moving down a concentration gradient), while large molecules, such as soluble proteins, are unable to cross the membrane.

◼ Chromatography

Most purification methods involve chromatography and involve a column of an insoluble material that can bind molecules based on specific properties common to proteins. The solution containing the mixture of proteins or coloured pigments is then allowed to pass through the column; the protein being studied may bind, while the impurities remain in solution and leave the column. The procedure is completed by eluting (removing) the proteins or pigments bound to the column.

The Bradford spectrophotometric assay

The Bradford spectrophotometric assay is used to measure protein concentration in solution. It is an indirect measurement of protein concentration, since it does not measure the light absorbed directly by the protein molecules in solution. The assay measures the light absorbed by a specific blue dye (Coomassie Blue R-250) that absorbs more of certain wavelengths of light (and less of others) when bound to proteins than when it is not.

> **Examiner guidance**
>
> A biochemical assay is a procedure for assessing or measuring the presence, amount or activity of a biochemical substance.

The increase in absorbance resulting from the interaction of the dye with protein is monitored using a spectrophotometer or colorimeter. To calibrate the absorbance to protein concentration, a series of protein standards are assayed and a plot of absorbance versus protein concentration (standard or calibration line) is generated. This graph is used to determine the concentration of proteins in experimental samples based upon the absorbance of those samples.

> **Expert tip**
>
> Some cuvettes are designed for visible light only. When the colorimeter is set for ultra-violet wavelengths (wavelengths of 340 nm or less) make sure that your cuvette does not have a large absorbance when it contains only water.

Concentration and dilution

◼ Units of concentration

The most common descriptions of concentration in biochemistry are molarity and percent-of-solute.

Molarity is defined as the amount (in mol) of solute per cubic decimetre (litre) of solution (denoted M). It is a widely used unit of concentration since the mole is directly related to the number of particles (usually ions or molecules).

> **Worked example**
>
> Calculate the mass of glucose ($C_6H_{12}O_6$) needed to make 50 cm^3 of a 5.0 mM solution.
>
> 180.18 g mol^{-1} × 0.05 dm^3 × 0.005 mol dm^{-3} = 0.045 g

> **Expert tip**
>
> The speed of dialysis can be increased not only by stirring the outside solution but also by increasing the surface area/volume ratio of the inside solution, as the rate of diffusion is proportional to the cross-section.

> **Expert tip**
>
> Ultra-violet/visible (UV/Vis) spectroscopy is an important technique for assaying, directly or indirectly, the concentration and concentration changes of various biomolecules. Using enzymes as an example, the concentration changes of substrate or product molecules in the presence of the enzyme provide important raw data about the activity of the enzyme.

> **Expert tip**
>
> Absorbance (*A*) is defined by the relationship: $A = \log_{10}(I_0/I)$. This is usually shown as A_x, where *x* = the wavelength of the radiation in nanometres; I_0 is the intensity of the incident radiation onto the sample and *I* is the intensity of the transmitted radiation. In addition: $A = \varepsilon l c$, where: ε = a constant for the absorbing substance (molar absorption coefficient, mol dm^{-3} cm), *l* = the length of the light path through the absorbing solution (typically 1 cm) and *c* = concentration of absorbing solution in mol dm^{-3}.

Molarity or molar concentration usually describes solutions of accurate concentration, where the molecular mass (molar mass) of the solute is known. Pure dry solids are weighed on an electronic analytical balance, and volumes are measured in volumetric flasks. In biochemistry, most concentrations are in the millimolar (mM), micromolar (μM) or nanomolar (nM) range.

Percent-of-solute is frequently used for liquids and solids of undetermined or variable molecular mass: volume per volume (most common) (v/v), weight per volume (w/v), and weight per weight (w/w) (where weight refers to mass).

A 6% (v/v) solution contains 6 cm^3 solute in 100 cm^3 total solution. A 6% (w/v) solution contains 6 g solute per 100 cm^3 total solution. A 6% (w/w) solution contains 6 g solute per 100 g total solution.

> ### Expert tip
>
> Sometimes these concentrations are not expressed as percentages, but rather simply as the number of mg (or μg) per cm^3 of solution.

> ### Worked example
>
> Calculate the concentration in % (w/w) for a 1.00 M solution of sodium chloride, NaCl (aq).
>
> $$\text{Concentration} = \frac{1 \text{ mol}}{1 \text{ dm}^3} = \frac{58.44\text{g}}{1 \text{ dm}^3} = \frac{5.844\text{g}}{100\text{cm}^3} = \frac{5.844\text{g}}{100\text{g}} = 5.8\% \text{ (w/w)}$$

▦ Dilutions

▦ Making volumes of specific concentrations

Many solutions used in biochemistry are prepared by the dilution of a more concentrated stock solution. A simple equation allows the dilution to be calculated readily:

$$c_1 \times V_1 = c_2 \times V_2$$

where c_1 is the concentration of the initial solution; V_1 is the volume of the initial solution available to be used for dilution (this may not be the total volume of the initial solution, and instead may be a small fraction of the initial solution), c_2 is the required final concentration and V_2 is the required final volume.

> ### Worked example
>
> You have a stock protein solution of 1000 μg cm^{-3} and you need 200 μl of 20 μg cm^{-3} protein solution. Deduce how the solution can be prepared.
>
> $$V_1 = V_2 \times \frac{c_2}{c_1}$$
>
> $$4 \text{ μL} = 200 \text{ μl} \times \frac{20 \text{ μg cm}^{-3}}{1000 \text{ μg cm}^{-3}}$$
>
> Hence you need to dilute 4 μl of the stock solution to a final volume of 200 μl (that is, by adding 196 μl of distilled water).
>
> If you wanted to make a solution of 1 μg cm^{-3}, the same equation would indicate that you need 0.2 μl of the 1000 μg cm^{-3} protein stock solution for 200 μl of the final diluted sample, but 0.2 μl is very difficult to measure accurately.
>
> You can change the final volume (that is, if V_2 is larger, then V_1 must also increase), or carry out serial dilutions (that is, instead of diluting the stock solution by a factor of 1000 in one step, dilute the stock solution, and then make a further dilution of the diluted stock).

■ ACTIVITY

10 From a stock solution of 100 mM ATP, calculate how much of this solution is added to distilled water to get 100 cm^3 of 15 mM ATP solution.

■ Simple dilution

A simple dilution is one in which a particular volume of a liquid or solution is mixed with a calculated volume of a solvent to form the required concentration. The dilution factor is the total number of unit volumes (cm^3, dm^3 and so on) in which your liquid or solution will be dissolved.

For example: a 1:10 dilution of glucose stock solution involves combining 1 unit volume (for example, 1 cm^3) of diluent (the liquid or solution to be diluted) with 9 unit volumes (eg 9 cm^3) of water. Hence, the dilution factor is 10, that is (1 + 9), and the glucose solution is now one tenth as concentrated or ×10 more dilute as it was before.

Expert tip

In a simple dilution, add one less unit volume of solvent than the desired dilution factor.

■ ACTIVITY

11 A 1:4 dilution of 50 mM Tris base is carried out. Calculate the final molar concentration in mol dm^{-3}.

■ Serial dilution

A serial dilution (Figure 2.22) is a series of simple dilutions which increase the dilution factor. The source of solution to be diluted for each step comes from the diluted solution of the previous dilution. In a serial dilution, the total dilution factor is the product of the individual dilution factors in each step up to it.

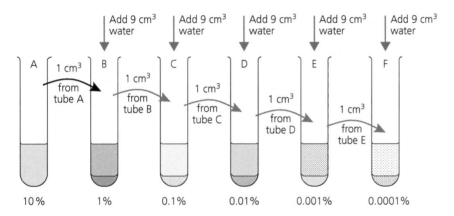

Figure 2.22 Making a serial dilution: dilution factor of 10 for each transfer

Final dilution factor = dilution factor 1 × dilution factor 2 × dilution factor 3 ... and so on.

Worked example
For example: A two-step 1:10^4 serial dilution of a bacterial culture.   The first step combines one unit volume (10 μL) with 99 unit volumes of culture (990 μL) resulting in a 1:100 dilution. In the second step, one unit volume of the 1:100 dilution is combined with 99 unit volumes of culture (990 μL) which gives a total dilution of 1:100 × 100 = 1:10000. The concentration of the diluted bacterial culture is now 10^4 less than in the original sample

Risk assessments for practical work

Examiner guidance

The three main parts of a risk assessment are:

- hazard identification – identifying safety and health hazards associated with laboratory work
- risk evaluation – assessing the risks involved
- risk control – using risk control measures to eliminate the hazards or reduce the risks.

Hazard labelling systems

The terms 'hazard' and 'risk' are frequently used interchangeably, but there is a distinct difference. A hazard is any source of potential damage or harm to an individual's health or life under certain conditions in the laboratory.

Risk is the chance or probability of a person being harmed or experiencing an adverse health effect if exposed to a hazard. Broken glass in the laboratory is a hazard and cuts are a potential source of harm. If broken glass is left lying around, the risk of harm is high; if it is cleared up immediately, using appropriate equipment, the risk of harm is low. Similarly, exposure of the skin to sodium hydroxide is a hazard and blistering of the skin is a potential source of harm. If you handle this chemical without gloves and safety goggles, the risk of harm is high. If you wear appropriate safety clothing, the risk of harm is reduced.

Risk assessment is the process of estimating the probability of harm from a hazard (the severity of the hazard multiplied by the probability of exposure to the hazard) by considering the process or the laboratory procedure that will be used with the hazard.

Carrying out a risk assessment involves estimating the risk and then identifying steps to minimize the risk: for example, reducing the quantity of the hazard being handled, using chemical fume hoods and protective barriers, such as plastic safety screens, devising safe procedures for handling the hazard, and using personal protective equipment, such as safety glasses and a laboratory coat.

Classifying hazardous chemicals

The Globally Harmonized System (GHS) is an internationally adopted system from the United States for the classification and labelling of hazardous chemicals. The GHS provides established descriptions and symbols (Figure 2.23) for each hazard class and each category within a class. This description includes a signal word (such as 'danger' or 'warning'), a symbol or pictogram (such as a flame within a red-bordered diamond), a hazard statement (such as 'causes serious eye damage') and precautionary statements for safely using the chemical.

> **Expert tip**
>
> Always read the label on a chemical reagent bottle to obtain and review basic safety information concerning the properties of a chemical. It is your responsibility in conjunction with your teacher to be fully aware of the hazards and risks of all chemicals you are using.

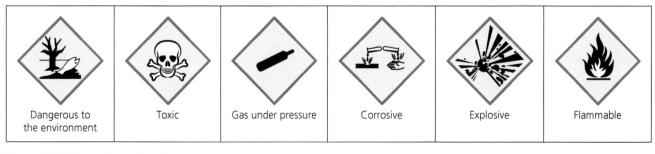

| Dangerous to the environment | Toxic | Gas under pressure | Corrosive | Explosive | Flammable |

Figure 2.23 Hazard warning signs

Safety Data Sheets

The Safety Data Sheet (SDS) is provided in the US by the manufacturer, distributor or importer of a chemical to provide information about the substance and its use. The SDS presents the information in a uniform manner. The information includes the properties of each chemical; the physical, health and environmental health hazards; protective measures; and safety precautions for handling, storing, disposing of and transporting the chemical.

Before an experiment

Carefully develop a list of all the chemicals used and the quantities needed in an experiment. You should always identify the substance you are working with and think about how you can minimize exposure to this in the experiment, considering the **exposure limit** for each substance or chemical.

Fully explain and demonstrate any new procedures or techniques that will be introduced in the investigation to your biology teacher.

> **Key definition**
>
> **Exposure limit** – this is the established concentration of a chemical that most people could be exposed to in a typical day without experiencing adverse effects. Exposure limits help in understanding the relative risks of chemicals.

Find and evaluate hazard information. This information is typically found on the SDS, which suppliers are required to provide the end-user. The label of the original container also contains valuable safety information.

Determine the minimum quantity of each chemical or solution that will be required for completion of an experiment. Build in a small excess, but avoid having large excesses that will require disposal.

Ensure that the proper concentrations are prepared and all chemical bottles are properly labelled: name, formula, concentration and any hazard warnings, such as corrosive or caustic, and its pictogram.

Examiner guidance

The source of heat for an experiment is an important consideration, particularly if any flammable solvents are used. Common laboratory hot plates are not designed for the heating of flammable or combustible chemicals. In no case should a burner be used to heat a flammable or combustible chemical. If flammable materials need to be heated, this should be done in small quantities in a hot water bath and in a fume hood. Never use a burner near a flammable substance.

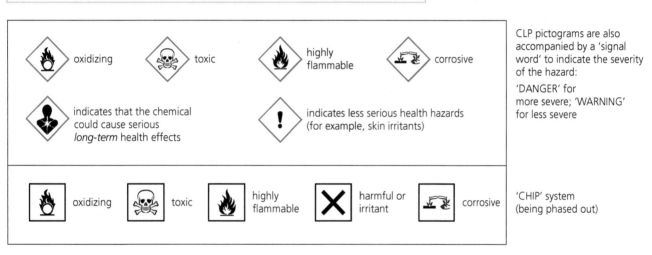

Figure 2.24 Comparing different hazard labels

Labels should be used to indicate the content in the disposal container. Write out all chemical names. If the content is a mixture of chemicals then indicate the major chemicals present and list the most hazardous chemicals.

Expert tip

When assessing safety, ethics and environmental issues, you should ensure that the following are considered and included in your IA report:

- evidence of a risk assessment
- an appreciation of the safe handling of chemicals or equipment (for example, the use of protective clothing and eye protection)
- the application of the International Baccalaureate animal experimentation policy
- a reasonable consumption of materials
- the use of consent forms in human physiology experimentation
- the correct disposal of waste
- attempts to minimize the impact of the investigation on field sites.

3 Cell structure and function practical skills

The effect of size and surface area on the movement of molecules

Equipment

- agar block containing phenolphthalein indicator, $20 \times 40 \times 10$ mm
- 5 test tubes in a test tube rack
- dilute hydrochloric acid (1.00 mol dm^{-3})
- stop clock
- ruler, sharp knife, white tile
- marker pen

Safety

- Phenolphthalein indicator can cause eye and skin irritation.
- Wear eye protection when handling the acid and indicator.
- Wash your hands after the experiment.

Information

This experiment explores the impact of the surface area : volume ratio on the rate of **diffusion** into cells.

In the experiment, the jelly blocks represent cells and the acid represents essential materials needed by cells for metabolic reactions. The acid diffuses into each of the blocks through the outside surfaces. The faster the substances reach all parts of a cell, the more likely the cell is to survive. The red dye in the agar blocks becomes pink and eventually colourless when mixed and reacted with the hydrogen ions in the acid. Acid moves into each of the blocks through the outside surfaces. Movement of the hydrogen ions, H^+(aq), is by simple diffusion.

1. Take a block of agar 40 mm $\times$ 20 mm $\times$ 10 mm. The jelly has been made with phenolphthalein indicator, which stains the jelly pink and turns colourless in the presence of acid.

2. Cut the agar block in half, to create two equal-sized blocks each $20 \times 20 \times 10$ mm. The blocks represent model cells, which will be used to investigate how quickly material can diffuse throughout the cell. Put one block aside as 'cell A'.

3. Cut the other block in half to create two blocks, each $20 \times 10 \times 10$ mm. Put one of these blocks aside as 'cell B'.

4. Cut the remaining block in half, so that you have two blocks, each $10 \times 10 \times 10$ mm. Put one of these aside as 'cell C'.

5. Cut the remaining block in half again, to make two blocks, each $5 \times 10 \times 10$ mm. Put one block aside as 'cell D'.

6. Cut the final piece in half, to give two blocks of $5 \times 5 \times 10$ mm. Put one of these aside as 'cell E' and discard the other. You now have five 'cells' in total, each of a different size.

7. Place each of the five model 'cells' in 1.00 mol dm^{-3} hydrochloric acid in a test tube, and start the stop clock. (Make sure you are wearing eye protection.) As the acid diffuses into the agar 'cells' the indicator will change from pink to colourless.

8. Observe what happens to the model cells. Record the time it takes for each cell to turn completely colourless. Compare the results for the different sized agar blocks.

9 Calculate the surface area, volume, and surface area : volume ratio for each agar block.

10 Repeat the experiment to improve reliability; take care when cutting the cubes to ensure your results are accurate. *How could you improve the precision of the experiment?*

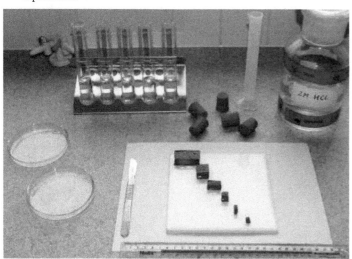

Figure 3.1 The 'jelly block' experiment

Results

Record your results in a table like Table 3.1.

Block	Time to turn colourless/s	Surface area of block/mm²	Volume of block/mm³	Surface area : volume ratio
A		(20 × 20) × 2 + (20 × 10) × 4 =	20 × 20 × 10 =	
B		(10 × 10) × 2 + (20 × 10) × 4 =	20 × 10 × 10 =	
C		(10 × 10) × 6 =	10 × 10 × 10 =	
D		(10 × 10) × 2 + (5 × 10) × 4 =	5 × 10 × 10 =	
E		(5 × 5) × 2 + (5 × 10) × 4 =	5 × 5 × 10 =	

Table 3.1 Example table of results for the jelly block experiment

Analysis

■ Hydrogen ions from the acid diffuse throughout the largest 'cell' in the longest period of time and the smallest in the shortest period of time.

■ The smaller cells have a larger surface area : volume ratio, meaning that acid diffuses into them more quickly.

■ If a graph is plotted, the time taken for the acid to move throughout the cell does not continue to decrease exponentially as cells get smaller, but begins to plateau. This means that there is an optimal size for cells, beneath which there is no benefit to a continued reduction in size.

■ Cells are small because this allows soluble substances to diffuse throughout them in a period of time that allows them to function and survive – in larger cells, diffusion takes too long to reach all parts of the structure.

Practical 1: Use of a light microscope to investigate the structure of cells and tissues, with drawing of cells

Aim

■ Use a light microscope to investigate the structure of cells and tissues.

■ Draw the cells you observe.

■ Calculate the magnification of drawings and the actual size of structures and ultrastructures shown in your drawings.

The ability to draw, label and annotate biological specimens is an important and useful biological skill.

■ Drawing biological specimens: general principles

Use a sharp pencil and draw clear, continuous lines. Do not use any form of shading. It might be helpful to use a magnifying glass and illumination (light). Include a title (with binomial name of species) and a scale.

If you are drawing from a microscope, state the combined magnification of the eyepiece plus objective lenses used when making the drawing, for example, ×100 (low power (Figure 3.3)) or ×400 (high power). Note that this is not the same as recording the scale.

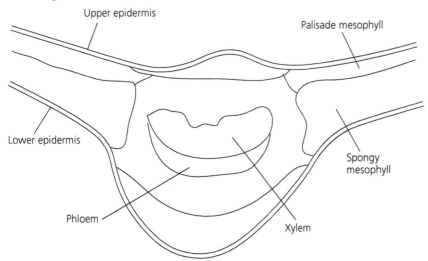

Figure 3.3 A low-power (¥100) drawing of a cross-section of a leaf of privet, *L. ovalifolium*

Rather than measuring the time for each block to lose its colour completely, you can measure the distance the acid has diffused in a specific amount of time and calculate the volume of the block that still shows red dye (volume = length × width × height). You can then calculate the volume into which the acid has diffused (diffused volume = total volume – volume that still has red dye). Finally, you can calculate the percentage of the total volume into which the acid has diffused: percentage diffused = 100 × (diffused volume/total volume).

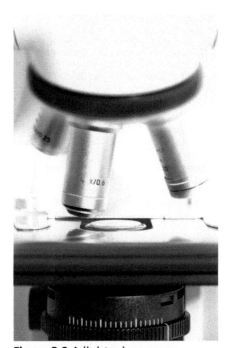

Figure 3.2 A light microscope

Expert tip

Scale bars can be used as a way of indicating actual sizes in drawings and micrographs, and can be used to calculate magnification. Magnification is calculated by dividing the actual length of the scale bar by the length indicated on the scale bar.

Living tissues, prepared for examination under the microscope, are typically cut into thin sections and stained. Stains are used to highlight structures in biological tissues for viewing. For example, acetic orcein can be used to stain genetic material (see page 49). A widely used staining method is the H&E system, which contains the two dyes: haematoxylin and eosin, where eosin stains structures, such as the cytoplasm in cells, pink or orange, and the haematoxylin stains nucleic acids a deep blue-purple colour.

Labelling

Use a sharp pencil and label all relevant structures, including tissues if a microscope is being used to view the specimen. Use a ruler to draw label lines and scale bars. Label lines should start exactly at the structure being labelled. Arrange label lines neatly and make sure they do not cross over each other. Labels should be written horizontally, not at the same angle as the label line. Add a scale bar immediately below the drawing if necessary.

Annotating

Annotation adds concise notes about the structures labelled on a biological drawing. It is often used to draw attention to structural or functional features of particular biological interest.

Scale and magnification

It is useful to give an indication of the scale or magnification of a biological drawing, especially for large specimens drawn without a microscope. For drawings made using microscopes, if the actual scale or magnification is not given, then indicate whether a low- or high-power lens was used, preferable stating the actual magnification achieved by the combined eyepiece and objective lens. Write this just below the title. See page 45 for further information about magnification and working out the actual size of biological objects.

Drawing from a microscope slide

The purpose of a low-power drawing is usually to show the distribution of the main tissues within an organ. No individual cells should be drawn. Identify and draw all the tissues and enclose each with lines. A representative portion can be drawn if the structure is symmetrical.

The purpose of high-power drawings (Figure 3.4) is to show as much accurate detail as microscopy will allow. If all the cells are the same type, then three or four cells might be sufficient to show both cell structure and the way in which the cells are arranged in relation to each other.

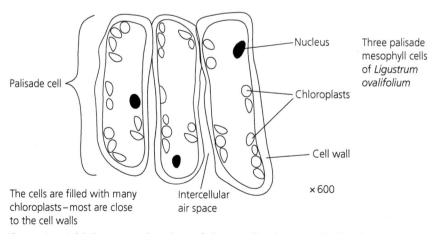

Figure 3.4 A high-power drawing of three palisade mesophyll cells

Calculating the actual size of specimens

Before you can measure the actual size of your specimen, you must calibrate the eyepiece graticule. Calibration allows you to work out the width (in μm) represented by one division of the eyepiece graticule.

Note: because the width represented by each division of the eyepiece graticule is different at different magnifications, the calibration must be done at each different magnification used. If only one magnification is used (for example, low power) then the calibration only needs to be done once.

Calibrating the eyepiece graticule

1 Insert a graticule into the eyepiece of the microscope by unscrewing the top lens and inserting an eyepiece that contains a graticule on the lower surface. (Alternatively, if you do not have eyepiece graticules, rest a free-standing graticule on the rim halfway down and replace the top lens.)

2 Place a stage micrometer slide on the stage of the microscope. Note that the smallest division on the stage micrometer equals 100 μm (this might vary between different designs of micrometer – check yours).

3 Using the low-power objective, focus the microscope on the stage micrometer. Rotate the eyepiece and move the slide to superimpose the scales of the eyepiece graticule and the stage micrometer (see Figure 3.5).

4 Count the number of divisions on the eyepiece graticule equivalent to 100 μm on the stage micrometer and, hence, calculate the length that one eyepiece division is equivalent to. For example, if three divisions are equal to 100 μm, then each division is equal to 33.3 μm at low power. Record your answer. This is the calibration factor.

5 You can now remove the stage micrometer and put your specimen on the slide. How many divisions on the eyepiece graticule does your specimen cover? Multiply the number of divisions by the calibration factor (ie if your specimen is four divisions across under low power then, using the calibration factor calculated above, the actual size of the specimen is 4 × 33.3 μm = 133.2 μm).

Figure 3.6 shows how this technique can be used to measure the size of a red blood cell.

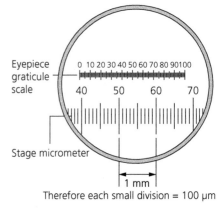

Figure 3.5 Use of stage micrometer and eyepiece graticule to calculate the size of objects under a microscope

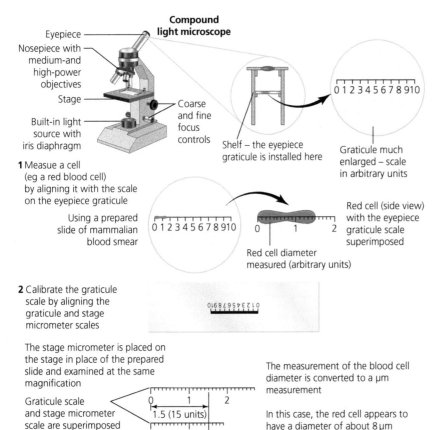

Figure 3.6 Measuring the size of cells

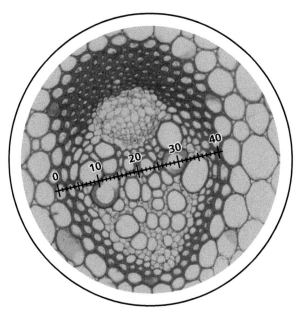

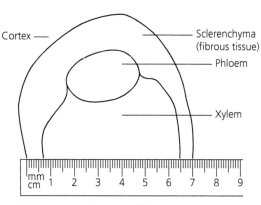

Distance across drawing = 70 mm
Distance across vascular bundle = 400 μm
Magnification = $7 \times 10^{-2} / 4 \times 10^{-4} = \times 175$

Figure 3.7 Using an eyepiece graticule to record small measurements from a microscope and calculate magnification of a plan drawing of a vascular bundle

■ Magnification

Magnification is the number of times larger an image is than the specimen. The magnification obtained with a compound microscope depends on which of the lenses you use. For example, using a ×10 eyepiece and a ×10 objective lens (medium power), the image is magnified ×100 (10 × 10). When you switch to the ×40 objective (high power) with the same eyepiece lens, the magnification becomes ×400 (10 × 40). These are the most likely orders of magnification used in your laboratory work.

You can determine the magnification by working out the actual size of the object you have drawn (for example, a plant cell) using an eyepiece graticule and stage micrometer (see above). You can calculate the magnification using the following formula:

$$\text{magnification} = \frac{\text{size of image}}{\text{size of specimen}}$$

For example, take a plant cell of 150 μm diameter, drawn by hand, where the drawing shows the cell at 15 cm diameter (150 000 μm). The magnification is 150 000/150 = 1 000 (that is, a magnification of ×1 000).

A memory diagram can be used to show how to calculate the magnification, actual size, or image size of an object (Figure 3.8). Remember the equation as *AIM* or *I AM*, and make sure you convert units so that they are the same for both *I* and *A*.

  I = size of image (drawing of an object on paper)

  A = actual size of the object being measured

  M = magnification (the size of an image compared to the actual size of the object, ie the number of times larger an image is than the specimen)

So, $M = \frac{I}{A}$; $A = \frac{I}{M}$ and $I = A \times M$.

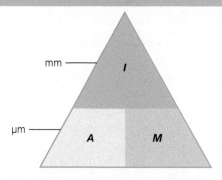

Figure 3.8 Memory diagram to help remember how to work out the magnification, actual size or image size of a specimen

■ **ACTIVITIES**

1 **Suggest** why it is useful to give an indication of the scale/magnification of a biological drawing.

2 The ability to draw, label and annotate biological specimens is an important and useful biological skill. **Outline** how you would draw a cross-section of a leaf of privet.

Practical 2: Estimation of osmolarity in tissues by bathing samples in hypotonic and hypertonic solutions

Aim

■ Put potato tissue in a range of sucrose solutions of different **osmolarity** to see how they change in mass and length.

■ Estimate the osmolarity of potato tissue by finding the concentration of sucrose where there is no change in mass or length.

■ Evaluate the experiment in order to comment on the accuracy of the results.

Equipment

■ 1.00 mol dm^{-3} sucrose solution

■ cork borer/chip-maker

■ distilled water

■ pipettes

■ burette, to deliver variable volumes

■ boiling tubes

■ electronic balances

■ stop clock

■ ruler or callipers

Safety

■ Take care cutting the potato chips. Use a white tile to cut the potato on and do not cut towards the body.

Procedure

1 Make up six sucrose solutions of 1.00, 0.80, 0.60, 0.40, 0.20 and 0.00 mol dm^{-3} (see Table 3.2).

2 Use the cork borer to prepare 30 chips of potato each 30 mm in length.

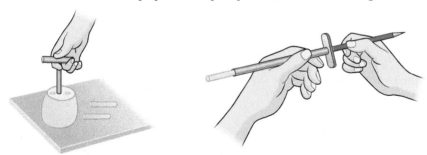

Figure 3.9 Preparing potato chips using a cork borer

3 Weigh and measure each chip and record the masses (each length should be 30 mm).

> **Key definitions**
>
> **Hypotonic** – when the external solution is less concentrated (has a lower solute potential) than the cell solution (cytosol), and there is a net inflow of water into the cell by **osmosis**.
>
> **Hypertonic** – when the external solution is more concentrated (has a higher solute potential) than the cell solution (cytosol), and there is a net flow of water out of the cell by osmosis.
>
> **Osmolarity** – the concentration of a solution expressed as the total number of solute particles per litre.

4   Put one chip in each of the solutions. Repeat the test five times with each sucrose concentration (ie have five boiling tubes containing the solution at each concentration, with a chip of known mass and length in each tube).

5   After 40 minutes, remove the chips and re-weigh and re-measure them – take care to remove any excess solution first (*why do you do this?*).

6   Calculate the percentage change in mass and the percentage change in length for each chip. This is calculated by working out the change in mass or length, dividing it by the original mass or length and multiplying by 100 (to produce a percentage).

7   Plot a graph of percentage change (*y*-axis) against sucrose concentration (*x*-axis) for both length and mass.

8   Estimate the concentration of the potato tissue (its solute potential). This is the point when there is no change in mass/length (ie no net osmosis because the solute potential is the same in the solution as the cell cytosol). *Is it the same for both length and mass?*

Volume of distilled water/cm^3	Volume of 1.00 mol dm^{-3} sucrose/cm^3	Concentration of sucrose/mol dm^{-3}
2.00	8.00	0.80
4.00	6.00	0.60
6.00	4.00	0.40
8.00	2.00	0.20

Table 3.2 Preparing different concentrations of sucrose solution, 0.80 mol dm^{-3} → 0.20 mol dm^{-3}

Experimental method

Accurate measurements are needed in scientific experiments. Accuracy relates to how close your results are to the true value. Accuracy can be improved by carefully measuring mass and length in this experiment. Limitations in your equipment and method will reduce the accuracy of the results. For example, in this experiment the solutions the potato is put in must be carefully prepared. If too much or too little water is mixed with the sucrose then the concentration will be incorrect.

Using a burette, rather than a pipette, to transfer the distilled water into the sucrose solution will improve accuracy because the water molecules will not stick to the sides of a burette as they might in a pipette.

Results

Record your results in a table. Table 3.3 shows you how to arrange your table.

Sucrose concentration/ mol dm^{-3}	Percent change (%) in the mass of the potato chip					
	Repeat 1	Repeat 2	Repeat 3	Repeat 4	Repeat 5	Mean
0.00						
0.20						
0.40						
0.60						
0.80						
1.00						

Table 3.3 A sample results table for an investigation into the effect of sucrose concentration on percent change in the mass of potato chips

Graph

Draw a graph of your results. The results should follow the pattern in Figure 3.10.

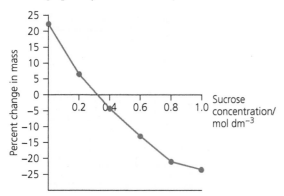

Figure 3.10 Percentage change in mass of potato chips at different sucrose molarities

Plot a graph for both percentage change in mass and percentage change in length.

Analysis

Describe and explain your results. The point where the line between the data points crosses the *x*-axis is the point at which there is no change in mass. This means that there is no net osmosis (*why is it incorrect to say that there is 'no osmosis'?*). This indicates the concentration of the cell solution (ie it is the same as the external solution), that is, the osmolarity.

Evaluation

1 What were the limitations of the experiment? How might they have affected the accuracy of the result? How could you improve the experiment to reduce these limitations and improve accuracy?

2 Comment on reliability, and identify any anomalous results. What is standard deviation, and what does this tell you about your results? (*Hint*: variation around the mean).

3 Which results provided the more accurate results – the percentage change in length or the percentage change in mass? Why is this?

The preparation of a root tip squash to demonstrate stages in mitosis

Aim

- To observe cells from an onion or garlic root that are actively dividing and going through the process of mitosis.

- To calculate the **mitotic index** and gain an understanding of the length of each stage of mitosis.

Equipment

- onion roots

- 10 cm^3 of a solution of 50:50 concentrated hydrochloric acid and absolute alcohol (ethanol)

- mounted needle

- teat pipette

- 10 cm^3 45% by volume glacial acetic acid (concentrated ethanoic acid)

- forceps

> **Key definition**
>
> **Mitotic index** – the number of cells undergoing mitosis divided by the total number of cells visible.

- acetic orcein in a bottle with dropper

- 50 ml beaker

- microscope with a light source

- microscope slides and cover slips

- stop clock

- scissors

- labels

- watch glasses

- lens tissue

- glass rod

- filter paper

Safety

- Concentrated hydrochloric acid and glacial acetic acid are corrosive: hydrochloric acid causes burns and the vapour from both irritates the lungs. Handle with care, and wear safety goggles and disposable gloves when using these chemicals.

- Acetic orcein is classified as low hazard but should still be used with caution, and skin contamination should be avoided.

- Wash hands after experiment.

- Absolute alcohol is very flammable and so should not be used near a naked flame, and the room should be well-ventilated.

Procedure

1 Take a few roots and cut off a few root tips, about 2 mm in length.

2 Using a mounted needle, place the root tips in the watch glass with the hydrochloric acid/alcohol mix. Leave for 10 minutes.

3 Transfer the root tips to another watch glass containing 45 % glacial acetic acid. Leave for 5 minutes.

4 Place a root tip on a microscope slide (clean the slide with a lens tissue if it is dirty).

5 Cover the tip with a drop of acetic orcein and squash the root with a vertical tapping motion using a flat-ended glass rod. Do not allow the material to dry out.

6 Put a cover slip on the slide and place between two layers of filter paper. Gently squash the preparation by pressing down on the cover slip through the layers of filter paper; squash firmly but carefully. Do not move the cover slip.

7 Examine under the microscope and identify cells that are dividing. *Can you see the different stages of mitosis?*

8 Draw, label and annotate two named stages of mitosis.

9 Now examine at least 10 different fields of view, and count at least 10 cells in each. Count the number of cells in each stage of mitosis and the total number of cells counted (the rest of the cells will be in interphase). Keep a tally of the total.

10 Calculate the mitotic index: this is the number of cells in mitosis divided by the total number of cells.

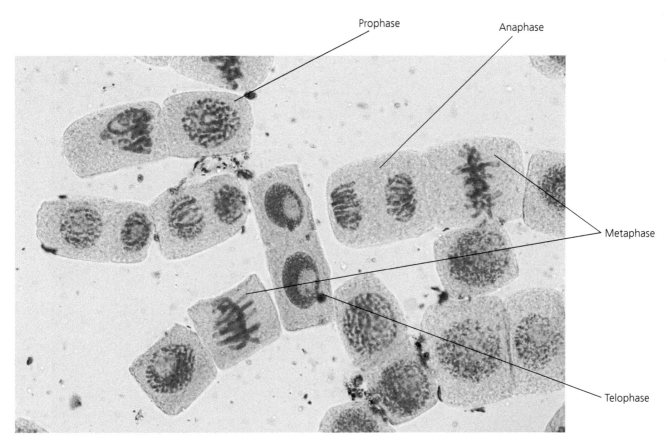

Figure 3.11 Root tip squash indicating the four mitotic stages

■ **ACTIVITY**

3 Analyse the following photomicrograph and calculate the mitotic index.

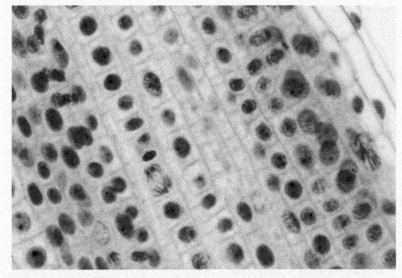

Figure 3.12 Root tip squash with cells undergoing mitosis

Physiology practical skills

Practical 6: Monitoring of ventilation in humans at rest and after mild and vigorous exercise

Aim

- Use a spirometer or data logger to find the **tidal volume** of lungs and calculate the **ventilation rate**.

- Interpret the traces resulting from the experiment.

- Investigate the effect of mild and vigorous exercise on the ventilation rate and the depth of breathing.

Advice

Ventilation can either be monitored by simple observation and simple apparatus or by data logging with a spirometer, or with a chest belt and pressure meter. You should measure ventilation rate and tidal volume, as these are included in the IB Biology syllabus. You might also want to measure vital capacity and residual volume (see Expert tip) to more fully understand ventilation.

Background information

A spirometer is used to measure the breathing rate and the capacity of the human lungs. A spirometer consists of a chamber suspended over water and counterbalanced so that, as air passes in and out, the chamber rises and falls. A slowly revolving drum (a kymograph) records the movement of the chamber. Alternatively, a motion sensor can be attached to the chamber which converts movement into electronic signals that can be interpreted by data-logging software. Another option is to use chest belts and pressure meters linked to a data logger. If neither spirometers nor data-logging equipment are available, simply counting the number of breaths before and after exercise will suffice.

Safety

- Do not use a spirometer unless supervised by a teacher.

- If you are using the spirometer and you experience any distress or difficulty breathing, remove the nose clip and mouthpiece and detach yourself from the machine immediately.

- Do not use the spirometer for more than a few minutes at a time.

- Please adhere to the IB ethical policy. All participants must be informed before commencing the investigation that they have the right to withdraw at any time. Pressure must not be placed on any individual participant to continue with the investigation beyond this point. Each participant must be informed of the aims and objectives of the research and in addition be shown the results of the research.

Variables

- Independent variable: type or intensity of exercise.

- Dependent variable: ventilation rate; volume of air inhaled/exhaled per minute; changes to tidal volume. Select one to measure for your experiment.

- Controlled variables: the way the exercise is carried out; the way that the dependent variable is measured; the length of time that the dependent variable is measured for; ensure ventilation rate and the depth of breathing have returned to the resting rate between measurements; the same person should carry out each investigation.

Key definitions

Tidal volume – the volume of air that a human breathes into and out of their lungs while at rest; this is around 500 cm^3, on average.

Ventilation rate – number of breaths per minute.

Expert tip

Vital capacity is the greatest volume of air that can be expelled from the lungs after taking the deepest possible breath. Residual volume is the volume of air still remaining in the lungs after the greatest possible forcible expiration – usually 980 to 1 640 cm^3.

Equipment

Figure 4.1 shows a spirometer being used to measure ventilation rate:

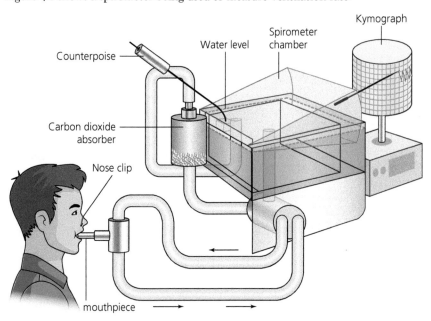

Figure 4.1 A spirometer

Expert tip

It is important to standardize the exercise in physiological experiments, for example, by running a fixed distance or using an exercise bike at a fixed setting.

Procedure

1   Set up the spirometer. The person who is the subject of the experiment puts on a nose clip and breathes into the mouthpiece (see above).

2   Collect spirometer traces (see Figure 4.2) from subjects before exercise, after light exercise (eg walking) and after more vigorous exercise (eg jogging/sprinting). For more quantitative data, the same activity could be done at different levels of intensity (eg running at different speeds on a running machine).

3   Repeat each experiment at least five times and calculate an average to make the experimental results more reliable, and to enable you to calculate the standard deviation (a statistical measurement of variation around the mean). *Ten repeats are ideally needed to ensure normal distribution of the data and to justify standard deviation measurements.*

4   Compare the rate of breathing and depth of breathing between traces (see Figure 4.2).

5   Describe the differences between the different traces. Explain what they show about human lung function and oxygen consumption.

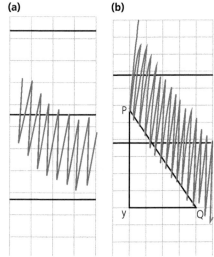

Figure 4.2 Spirometer traces

Examine the two traces in Figure 4.2: trace (a) taken after light exercise and (b) after vigorous exercise. Both traces have declining slopes as the spirometer removes carbon dioxide from the air breathed out (for studies investigating metabolic activity) and so the total volume of gases decreases over time.

Breathing rate can be calculated by measuring the number of peaks in the spirometer trace per minute. The depth of breathing can be calculated by measuring the average height of each peak, from the bottom of the peak (person has breathed in) to the top (person has breathed out).

Analysis

Changes in ventilation depth and rate at higher levels of activity occur because:

■   increased muscular contractions require more energy

■   increased aerobic **respiration** provides the increased energy needed

■   aerobic respiration uses oxygen, and so increased activity leads to increased demand for oxygen

■   increased demand for oxygen leads to increased depth and rate of breathing.

Ideas for investigations

- The body's demands on the circulatory system change and, as a consequence, the heart rate is constantly adjusted. The heart may beat between 50 and 200 times a minute. The effect of regular exercise on heart rate can be easily measured. Recovery rate can also be measured. Heart rate can be measured directly using a stethoscope, or by taking a pulse. Ventricular contractions force a wave of blood through the arteries, and the expansion of the arteries can be felt as a pulse, particularly where the artery is near the skin surface and passes over a bone. The pulse is traditionally taken above the wrist. The measurement of a pulse can be improved using digital pulse sensors, and many mobile phones have an app that can measure heart rate.

- Other human physiology investigations include change in reaction time to audio and visual stimuli. Reaction time can be measured using the 'ruler drop test' – a metre rule is held by the investigator so one end (the 10 cm mark) is situated between the thumb and forefinger of the individual undergoing the test. The ruler is dropped without warning and the subject of the investigation catches the ruler as quickly as possible. The distance the ruler has dropped is converted to time using the formula: $t = \sqrt{\frac{2s}{g}}$, where t = time (seconds), s = distance dropped (metres), g = 9.81 m s^{-2} (acceleration due to gravity).

- Computer programs can be used to carry out reaction time tests. For example:

  https://faculty.washington.edu/chudler/java/redgreen.html

  http://www.freewebarcade.com/game/sheep-dash/

  http://cognitivefun.net/test/16

  The independent variable in these tests could be, for example, age, time of day, gender.

In these experiments, 20–25 individuals should be used so that sufficient replicate data are obtained. As far as possible, other variables should be controlled (such as age and general fitness). Confounding variables should be monitored (see page ix) and recorded.

Practical 7 (*AHL only*): Measurement of transpiration rates using potometers

◼ Using a potometer

Aims

◼ Understand how a potometer works.

◼ Set up a potometer to measure **transpiration** rate.

Background information

The rate of transpiration can be measured in the lab using a potometer (literally a 'drinking meter'). The apparatus consists of a leafy shoot inserted into a tube (the seal must be airtight) which is attached to a capillary tube. The capillary is attached to a reservoir of water. The apparatus must be set up underwater to ensure that the column of water is continuous between the plant and the capillary. An air bubble is introduced into the capillary. As water transpires from the leaf, water is pulled up the tube and along the capillary, so the rate of movement of the air bubble (ie the distance moved per minute) can be used as a measure of transpiration rate (although the potometer actually measures water uptake rather than water loss – see comments below). The tap below the reservoir allows the bubble to be reset so that a new measurement can be made.

A potometer (see Figure 4.3) does not measure transpiration directly (if it did, it would more likely be called a 'transpirometer'): it actually measures the rate of water uptake by the cut stem. The rate of uptake of water will not be the same as the rate of transpiration: some of the water (around 5 %) remains in the plant for photosynthesis and to keep the cells turgid, and it is the remaining water (around 95 %) that is lost from the plant through transpiration. The difference between water uptake and water loss can be important for a large tree, but for a small shoot in a potometer the difference is usually negligible and can be ignored.

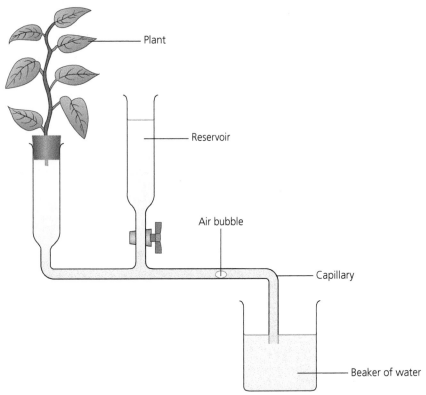

Figure 4.3 A potometer

Design an experiment to test hypotheses about the effect of temperature or humidity on transpiration rates

Background information

The potometer can be used to investigate how various environmental factors affect the rate of transpiration. Factors affecting transpiration include:

- Light: light stimulates the stomata to open allowing gas exchange for photosynthesis. This process also increases transpiration because the open stomata allow water to diffuse from the leaf. This is a problem for some plants as they can lose excessive amounts of water during the day and wilt.

- Temperature: high temperature increases the rate of evaporation of water from the spongy mesophyll cells and reduces air humidity, so transpiration increases.

- Humidity: air inside the leaf is saturated (relative humidity = 100 %): the lower the relative humidity outside the leaf the faster the rate of transpiration as the concentration gradient of water vapour is steeper. High humidity means a higher water potential in the air, so a lower water potential gradient between the leaf and the air, and so less evaporation.

- Air movements: wind blows away saturated air from around stomata, replacing it with drier air, so increasing the water potential gradient and increasing transpiration.

As well as these environmental/extrinsic factors that affect the rate of transpiration, there are also intrinsic factors which depend on the leaf structure. Intrinsic factors affecting the rate of transpiration include:

- leaf surface area

- thickness of epidermis and cuticle

- frequency of stomata/size of stomata/position of stomata.

Procedure

1 Choose an extrinsic abiotic variable to change and decide how you will vary this factor.

2 You must decide how many results you need.

3 How will you keep other variables (the controlled variables) the same? Why do you need to do this?

4 Set up a potometer and carry out the experiment.

5 Plot a graph of your results: independent variable (the abiotic factor you are changing) on the x-axis and dependent variable (the variable you are measuring, ie the distance the bubble has moved) on the y-axis. Transpiration rate is shown by the gradient of the line you plot.

6 Describe and explain your results.

7 Write your conclusion.

8 Evaluate your experiment.

As well as rate of movement of the air bubble, it is also possible to calculate the volume of water taken up by the plant. The volume of water taken up can be calculated using the following formula:

volume of water $= \pi r^2 \times$ distance

where r is the radius of the capillary tube (usually around 0.5 mm).

Vernier callipers or a micrometer screw gauge can be used to precisely measure the width of the capillary tube and the distance moved. It is, therefore, possible to calculate the total volume of water absorbed by the plant per minute for each of the environmental conditions you measure.

■ Suggested experiments to investigate the effect of an abiotic variable on the rate of transpiration

Suggested experiments

■ Air movement: use a fan (either have the fan on or off).

■ Temperature: use a fan with variable temperature (eg a hairdryer); either have hot wind or cold wind (note that the wind speed must be constant, with only the temperature varying).

■ Light: lamp on or off, in a darkened room.

■ Humidity: put a large transparent plastic bag over the plant to create a humid environment.

Procedure

Carry out your experiment and plot your results.

■ Take care to measure from the left side of the bubble (not the right).

■ Record the distance moved by the bubble every minute.

■ Repeat each test five times and take an average of your results.

Analysis

Use your knowledge of water concentration gradients between the inside of the leaf and the outside air, and (if relevant) how temperature increases the rate of evaporation of water, to fully explain your results. Figure 4.4 might help you.

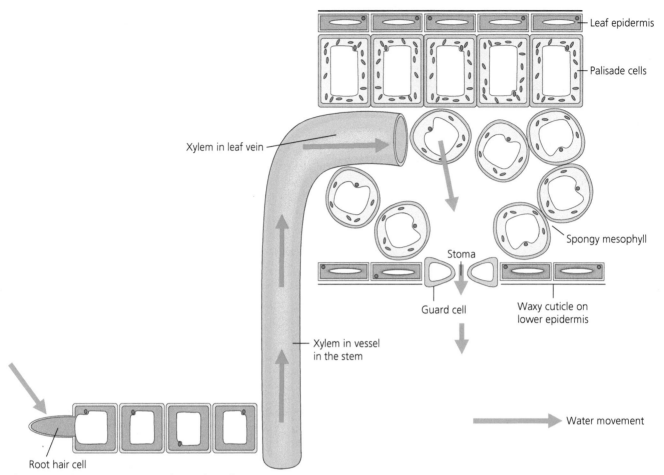

Leaf epidermis

Palisade cells

Xylem in leaf vein

Spongy mesophyll

Stoma

Guard cell

Waxy cuticle on lower epidermis

Xylem in vessel in the stem

Water movement

Root hair cell

Figure 4.4 Water movement through a plant

■ ACTIVITIES

In an investigation of water absorption and transpiration in a plant over a 24-hour period, readings were obtained at 4-hour intervals and were related to the light intensity. The results are shown in the table.

Time	Light intensity as a % of maximum	Water absorbed in 4 hours/g	Water transpired in 4 hours/g
04:00	0	6	1
08:00	70	6	8
12:00	100	14	20
16:00	100	22	29
20:00	10	13	10
24:00	0	8	3

1 **Present** these results graphically in a way that makes clear the pattern of water uptake and loss of water vapour during the experiment.

2 **Describe** the changing pattern of water absorption and transpiration in the plant during the 24-hour period.

3 **Explain** how the processes of water absorption and water loss are related in the intact plant.

Ideas for investigations

A potometer could be used to study different species of plant, such as those growing in woodland compared to those in more open habitats. Leaf sections from the plants can be taken and examined under the microscope – differences in leaf morphology can be linked with differences in potometer readings.

Expert tip

Potometer experiments can also be carried out by following mass loss on an electronic balance, if other equipment shown in Figure 4.3 is not available. Add a layer of oil to the water to prevent evaporative loss, so that mass loss is only due to transpiration.

Expert tip

Using a potometer to investigate the effects of abiotic variables on water uptake in plants is one of the prescribed practicals in your IB Biology syllabus (see page xii in the Introduction). Your IA cannot directly replicate one of these experiments. You can use the basic experimental procedure, but must modify it in some way so that it is adapted to a distinctive research question.

Figure 4.5 shows a section through one of the leaves of a plant that floats on top of water.

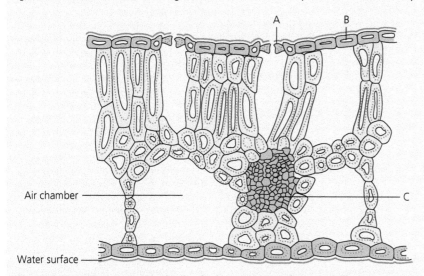

A   B

Air chamber

C

Water surface

Figure 4.5 Section through the leaf of an aquatic plant that floats on the surface of the water

4   **Identify** structures **A**, **B** and **C**.
5   **Suggest** a function of the air chambers in Figure 4.5 related to the maintenance of the position of the leaf.
6   **State** a different function of the air chambers related to photosynthesis.
7   **State** three ways in which the structure of this leaf differs from that of a typical plant growing on land.

Investigating respiration

The results from experiments involving measurement of respiration rates in germinating seeds or invertebrates can be analysed by using a respirometer.

The rate of respiration of an organism is an indication of its demand for energy. Respiration rate, the uptake of oxygen per unit time, can be measured by means of a respirometer (Figure 4.6). The manometer in this apparatus detects changes in pressure or volume of a gas. Respiration by tiny organisms (germinating seeds or fly maggots are ideal, and grasshoppers can also be used) that are trapped in the chamber of the respirometer alters the composition of the gas there, once the screw clip has been closed.

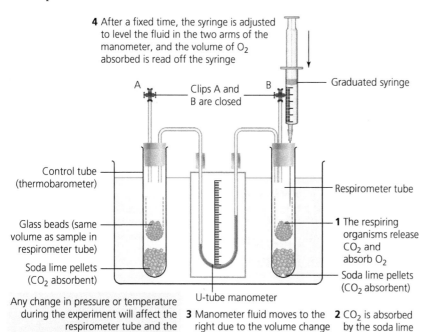

4 After a fixed time, the syringe is adjusted to level the fluid in the two arms of the manometer, and the volume of O_2 absorbed is read off the syringe

A   Clips A and B are closed   B   Graduated syringe

Control tube (thermobarometer)

Respirometer tube

Glass beads (same volume as sample in respirometer tube)

1 The respiring organisms release CO_2 and absorb O_2

Soda lime pellets (CO_2 absorbent)

Soda lime pellets (CO_2 absorbent)

Any change in pressure or temperature during the experiment will affect the respirometer tube and the thermobarometer tube equally.

U-tube manometer

3 Manometer fluid moves to the right due to the volume change caused by uptake of oxygen

2 CO_2 is absorbed by the soda lime

Figure 4.6 A respirometer to measure respiration rate

Common mistake

Some students have the misconception that plants undergo photosynthesis only and animals only undergo cellular respiration. However, plant cells also contain mitochondria and are continually respiring.

Expert tip

There are many simple respirometers that can be used. In each case, an alkali is used to absorb and react with CO_2 (it is neutralized), so reductions in volume are due to oxygen use. Temperature should be kept constant to avoid volume changes due to temperature fluctuations.

Respirometers can be used to find the volume of oxygen taken in by an organism and the volume of carbon dioxide produced.

Oxygen taken in

■ The organism is placed in a gauze basket positioned above soda lime (a mixture of calcium hydroxide and sodium hydroxide or potassium hydroxide) in a boiling tube (respirometer tube in Figure 4.6).

■ Soda lime absorbs carbon dioxide.

■ The boiling tube is airtight and connected to a capillary tube that has a scale attached (U-tube manometer in Figure 4.6).

■ The apparatus contains air, sealed by liquid in the capillary tube.

■ The organism takes in oxygen during respiration, and the volume of air in the boiling tube decreases.

■ Carbon dioxide produced by the organism does not replace the oxygen used, as it is absorbed by the soda lime.

■ The volume of air in the boiling tube decreases in proportion to the oxygen taken in, and the pressure falls.

■ Coloured water is drawn up the capillary tube.

   □ The distance moved by the coloured water indicates the volume of oxygen taken in.

   □ The volume of oxygen taken in per minute can also be calculated (see Expert tip).

Carbon dioxide produced

■ Remove the soda lime from the apparatus (Figure 4.6).

■ The difference between the distance the water moved with and without the soda lime indicates the volume of carbon dioxide produced.

■ Without soda lime, carbon dioxide is not removed and the water might not move at all: if the same volume of oxygen is taken in as carbon dioxide is produced, there will be no change in the volume of gas and the water will not move.

■ If less carbon dioxide is produced than oxygen is taken in, the fall in pressure will move the water a little towards the organism.

The resulting reduction in the volume of air in the respirometer tube in a given time period can be estimated. This is equivalent to the volume of air from the syringe that must be injected back into the respirometer tube to make the manometric fluid level in the two arms equal again. Where the respirometer tube contains soda lime, this volume is equivalent to the volume of oxygen taken up by the respiring organisms.

Ethical assessment

The use of invertebrates in respirometer experiments has ethical implications. IBO guidelines state that '*Any experimentation should not result in any pain or undue stress on any animal (vertebrate or invertebrate) or compromise its health in any way*'. Animals must be handled with care and must not be exposed directly to soda lime. They should only be used for brief periods of time, and returned to a safe environment once the experiment is completed.

■ Measuring the respiratory quotient of small animals

Respiratory quotient (RQ) is the volume of carbon dioxide produced divided by the volume of oxygen used:

$$RQ = \frac{\text{carbon dioxide produced}}{\text{oxygen taken in}}$$

Expert tip

• The soda lime removes the carbon dioxide gas released by the respiring organisms.

• Any change in the volume of gas, and therefore pressure, in the glass tube containing the organisms is therefore due to oxygen uptake by the respiring organisms.

• As a result, the coloured liquid in the attached capillary tube moves towards the respirometer tube.

Key definition

Respiratory quotient – ratio of the volume of carbon dioxide produced to the volume of oxygen used in respiration.

The importance of the RQ is that it can tell you what kind of substance is being oxidized by the animal (or plant), ie the substrate being used in respiration. Theoretical RQs for the complete oxidation of carbohydrate, fat and protein are:

Substrate	RQ
Carbohydrate	1.0
Fat	0.7
Protein	0.9

Table 4.1 Respiratory quotients (RQ) for a range of different biological molecules. RQs for fats and proteins are variable depending on the specific molecule being respired.

Procedure

1 Set up a respirometer as in Figure 4.6.

2 Mark with a black pen, and record, the height of the coloured water.

3 Put some woodlice into the specimen chamber, on the perforated zinc tray. **HANDLE WITH CARE**.

4 Leave for 5 minutes and then measure the height of the coloured water (in mm) above the starting point (the black line you drew in step 2).

5 Repeat the experiment, ideally five times. Reset the equipment by letting air into the boiling tube.

6 Work out the average distance the water has moved per minute.

7 Record the mass of your woodlice using an electronic balance.

8 Work out the oxygen uptake per gram of woodlice, using the following method:

Knowing the distance moved by the water and the diameter of the capillary tube, calculate the volume of oxygen using the formula:

Volume of oxygen $= \pi r^2 l$ (where l = distance moved by water)

Knowing the mass of the woodlice, calculate the oxygen uptake per gram of woodlice.

For example:

$$\frac{\text{total volume of oxygen taken in (mm}^3)}{\text{total mass of woodlice (g)}} = \frac{50}{10} = 5 \text{ mm}^3 \text{ oxygen per gram of woodlice}$$

9 Remove the soda lime from the boiling tube and repeat the experiment (steps 4–6 above).

10 Calculate the volume of carbon dioxide produced in the following way:

■ Work out the difference in distance travelled between the experiment with soda lime and the one without it.

(distance travelled experiment A – distance travelled experiment B)

■ Use this measurement of the distance moved due to carbon dioxide to calculate the volume of carbon dioxide released per gram of woodlice (following the method outlined above). (Note: *if the water has not moved, the volume of carbon dioxide will be the same as that of oxygen*).

11 Work out the respiratory quotient using this equation:

$$RQ = \frac{\text{carbon dioxide produced}}{\text{oxygen taken in}}$$

12 Repeat the experiment using a different animal (maggot or stick insect). Do you get a similar or different RQ? What does this tell you about the metabolism of the two animals?

For a hexose sugar such as glucose, the equation for complete oxidation is:

$$C_6H_{12}O_6 + 6O_2 \rightarrow 6CO_2 + 6H_2O$$

The RQ is hence:

$$\frac{6CO_2}{6O_2} = 1.0$$

In fats, the ratio of oxygen to carbon in the molecule is far smaller than in a carbohydrate. A fat therefore requires a greater quantity of oxygen for its complete oxidation and thus has an RQ less than one:
The equation for complete oxidation of stearic acid (octadecanoic acid) is:

$$C_{18}H_{36}O_2 + 26O_2 \rightarrow 18CO_2 + 18H_2O$$

The RQ is hence:

$$RQ = \frac{18CO_2}{26O_2} = 0.7$$

Analysis of results

In theory, we might expect an organism to have one of the three RQs in Table 4.1 (page 59), or a close approximation to it, depending on the type of food being respired. However, this rarely happens in practice. A respiratory substrate might not be oxidized completely, and often mixtures of substrates are used in the body. Most animals have a resting RQ between 0.8 and 0.9. The human body generally has an RQ around 0.85. As protein is not normally used to a great extent for respiration (except during starvation), an RQ of slightly less than 1.0 can be taken to mean that fat and carbohydrate are being respired.

■ A high RQ is usually the result of conversion of carbohydrate to fat, as carbon dioxide is liberated in this process. This is most noticeable in organisms that are laying down extensive fat reserves.

■ A low RQ indicates that some (or all) of the carbon dioxide released in respiration is being used by the organism (plants – photosynthesis; animals – construction of calcareous shells).

■ **ACTIVITIES**

8   In the respirometer (Figure 4.6), **explain** how changes in temperature or pressure in the external environment are prevented from interfering with measurement of oxygen uptake by respiring organisms in the apparatus.

9   The experiment shown in Figure 4.6 was repeated with maggot fly larvae in tube B, first with soda lime present and then with water in place of the soda lime. The volume change with soda lime was 30 mm³ h⁻¹, but without soda lime it was 3 mm³ h⁻¹. **Analyse** these results, explaining the significance of each value.

10 **Evaluate** the use of invertebrates in respirometer experiments and its ethical implications.

Investigating photosynthesis

■ **ACTIVITY**

11 Consider an investigation involving photosynthesis. Add additional examples of variables to a table like the one below.

Environmental variables	Plant or leaf variables
Light intensity (level of brightness)	Leaf colour (chlorophyll concentration)

■ Using floating leaf discs for investigating photosynthesis

Leaves, when placed in water, usually float. This is because air within the spongy mesophyll tissue reduces the density of the leaf.

This experiment uses small discs of leaves, cut using a cork borer or straw (see Activity below). The air within the air-spaces is removed by placing the leaf discs in a syringe and submerging them in water under pressure (Figure 4.7). Once all the leaves have sunk to the bottom of the syringe, the leaves can be placed under different light intensities, or light spectra (wavelengths of visible light), to investigate how these variables affect the rate of photosynthesis – as the leaf cells photosynthesize, they produce oxygen, which increases the buoyancy of the tissue and causes the leaf discs to float. The faster the leaf discs rise to the surface, the faster the rate of photosynthesis.

Baking soda can be used as an alternative to sodium hydrogencarbonate solution. The hydrogencarbonate ion provides a source of dissolved carbon dioxide for photosynthesis. One-eighth of a teaspoon of baking soda should be mixed in 300 ml of distilled or cooled, pre-boiled water; 1 drop of dishwashing liquid can be added to allow the hydrophobic surface of the leaf to become wet and allow the solution to be drawn into the leaf. A small quantity of detergent should be added, so that no foam is created. If the solution generates soapsuds then it should be diluted with more water and hydrogencarbonate solution.

Equipment

- cork borer
- syringe
- sodium hydrogencarbonate solution (or baking soda – see Expert tip)
- leaves from a broad-leaved plant such as the common geranium (*Pelargonium hortorum*)
- distilled water, or cooled, pre-boiled water

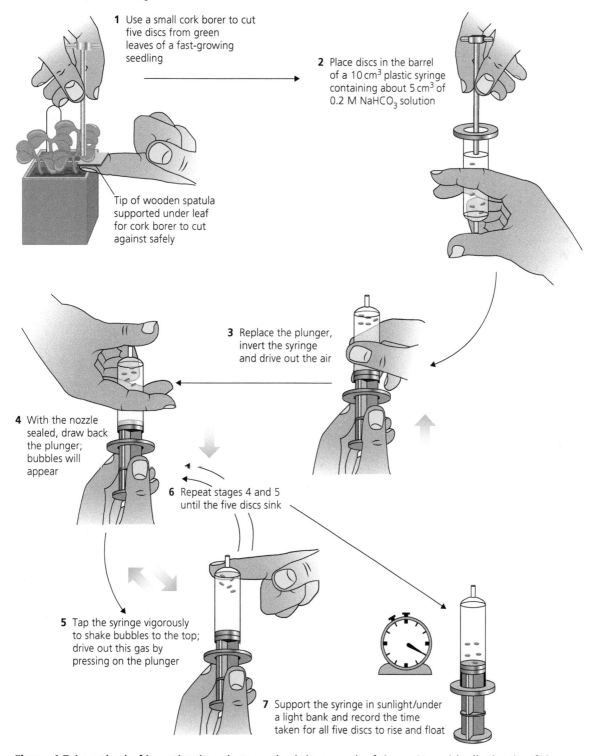

1 Use a small cork borer to cut five discs from green leaves of a fast-growing seedling

Tip of wooden spatula supported under leaf for cork borer to cut against safely

2 Place discs in the barrel of a 10 cm³ plastic syringe containing about 5 cm³ of 0.2 M NaHCO₃ solution

3 Replace the plunger, invert the syringe and drive out the air

4 With the nozzle sealed, draw back the plunger; bubbles will appear

6 Repeat stages 4 and 5 until the five discs sink

5 Tap the syringe vigorously to shake bubbles to the top; drive out this gas by pressing on the plunger

7 Support the syringe in sunlight/under a light bank and record the time taken for all five discs to rise and float

Figure 4.7 A method of investigating photosynthesis in green leaf tissue. Note: ideally the tip of the syringe should not be placed directly against your bare thumb, but rather against a barrier of parafilm or other impermeable material.

Once all the discs have floated to the top, place the syringe in a dark room or cupboard and record the time taken for the discs to sink back to the bottom (the oxygen bubbles will be used up during respiration, and so the buoyancy of the leaves will decrease).

Ideas for investigations

As an alternative to light intensity, light can be filtered through coloured filters so the leaves are exposed to different parts of the visible spectrum. Investigate whether there is a difference in the time it takes for them to float. Under which parts of the spectrum is the rate of photosynthesis fastest? Or the slowest? Relate your results to the absorption and action spectra.

■ ACTIVITIES

Many plants produce two types of leaves. One type is produced where the leaves develop in full sunlight and are called 'sun leaves'. The other type is produced where the leaves develop in the shade and are called 'shade leaves'.

Figure 4.8 shows the two different types of leaves collected from the same plant.

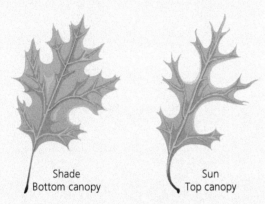

Shade
Bottom canopy

Sun
Top canopy

Figure 4.8 Two leaves collected from the same plant – one from a shady site and one from the top canopy where sunshine is abundant

A student investigated photosynthesis in both types of leaf using leaf discs. Figure 4.9 shows the method used by the student to obtain the leaf discs.

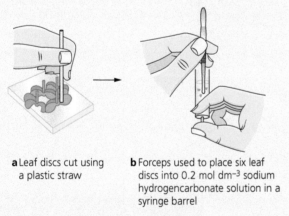

a Leaf discs cut using a plastic straw

b Forceps used to place six leaf discs into 0.2 mol dm^{-3} sodium hydrogencarbonate solution in a syringe barrel

Figure 4.9 Extracting leaf discs from a plant

The student then carried out the following actions:

c replaced the plunger into the syringe, turned the syringe upside down and pushed in the plunger to force out all the air

d placed a finger over the open end of the syringe and pulled down the plunger to create a vacuum

e tapped the side of the syringe to remove air bubbles

f repeated steps **c–e** until all the leaf discs sank to the bottom of the syringe.

Figure 4.10 shows the appearance of the leaf discs in the syringe as they photosynthesize.

12 State the purpose of the sodium hydrogencarbonate solution.

13 Explain why the leaf discs rise as photosynthesis proceeds.

The following table shows the results obtained from the investigation.

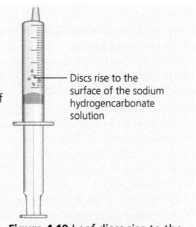

Discs rise to the surface of the sodium hydrogencarbonate solution

Figure 4.10 Leaf discs rise to the surface when exposed to light

| *x*-axis | Time taken for all six leaf discs to float/s (± 0.5 s) | |
	Sun leaves	Shade leaves
500	24.0	18.0
1 000	18.0	14.0
1 500	14.0	12.0
2 000	10.0	12.0
2 500	8.0	12.0
3 000	8.0	12.0

14 Describe how the time taken for the leaf discs to float is related to the rate of photosynthesis.

15 Plot the information in the table (with a suitable label for the *x*-axis) and state two conclusions that can be made about photosynthesis in sun leaves and shade leaves.

16 Suggest an advantage to tall trees of having both sun leaves and shade leaves.

■ Investigating how light intensity affects the rate of photosynthesis

Temperature, light intensity and carbon dioxide concentration are possible **limiting factors** on the rate of photosynthesis. This investigation uses an aquatic plant, pondweed (eg *Elodea canadensis*, *Elodea nuttallii*, or *Cabomba* spp.) to measure the effects of changing light intensity on the rate of photosynthesis.

Expert tip

Light intensity is a property of the surface emitting the light. Therefore, light intensity actually refers to the amount of energy transmitted. More accurately in such experiments, the irradiance level should be referred to rather than light intensity. The irradiance level is the level of illumination obtained, ie the amount of light actually striking the surface of objects (leaves). This depends upon the intensity of, and distance from, the light source.

The rate of photosynthesis will depend upon irradiance, the amount of light striking the leaf, rather than light intensity, the brightness of the bulb. A 60 watt light placed very close to a leaf will provide more light for photosynthesis than a 100 watt bulb located much further away.

Irradiance also encompasses units of area (square metre, m^2) and time (min), thus giving the amount of light energy striking a 2-dimensional surface over a period of time.

- Research question: how does light intensity affect the rate of photosynthesis?
- Independent variable: distance of pondweed from lamp (light intensity can be assumed to be proportional to the inverse square of the distance, or measured directly with a lux meter).

Key definition

Limiting factor – factor present in the environment in such short supply that it restricts life processes; for example, it may be a variable that restricts the rate of photosynthesis.

- Dependent variable: number of bubbles of oxygen produced per minute.

- Controlled variables: temperature; concentration of carbon dioxide (controlled by adding sodium hydrogencarbonate to the water); surface area of plant (controlled by using the same plant).

- Hypothesis: as the distance from the lamp decreases, the number of bubbles of oxygen produced per minute increases.

Procedure

Set up the apparatus shown in Figure 4.11.

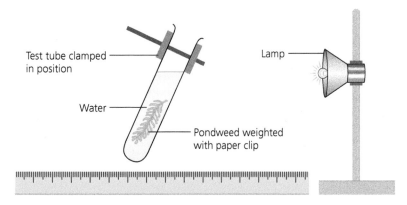

Figure 4.11 Measuring the effect of distance of a light source on the number of bubbles produced per minute by pondweed

1 Take a test tube and two-thirds fill it with 1 % sodium hydrogencarbonate solution. The concentration of sodium hydrogencarbonate controls the quantity of dissolved carbon dioxide in the water.

2 Take a piece of pondweed (eg *Elodea*) and cut the bottom of the stem at a 45° angle (so that oxygen bubbles will be readily released from the stem in water).

3 Put the pondweed in the test tube so that the cut stem is uppermost. Make sure the cut end is not touching the side of the test tube. Add a paperclip to the bottom end of the pondweed to weigh it down so it does not float to the surface.

4 Place a desk lamp at a measured distance from the pondweed, eg 5 cm. Let the plant acclimatize for one minute, until a regular stream of bubbles is given off. Then record the number of bubbles produced in one minute.

5 Move the lamp to a further fixed distance from the plant, eg 10 cm, and repeat the procedure.

6 Continue the experiment until you have measured the number of bubbles released per minute for at least 10 distances.

7 Repeat each experiment three times to ensure that you have reliable, valid quantitative results.

8 Once you have correctly collected, organized, transformed and presented your data in a table, plot a graph of your mean results.

9 Apply a curve of best fit to your mean data.

10 Accurately interpret your data and explain your results using correct scientific reasoning.

11 Evaluate the validity of the hypothesis based on the outcome of the investigation.

12 Evaluate the validity of the method and results based on the outcome of the investigation.

13 Explain improvements or extensions to the method that would benefit the investigation. For example, the experiment has several assumptions, eg that bubble size is constant, although in practice bubble volume varies, and so measurement of the volume of oxygen released would be an improvement to the method.

Safety

- Keep test tubes in racks, not loose on the bench as they might fall on the floor and break.

- Take care when cutting the plant if you are using sharp scissors.

■ Extension

The equipment shown in Figure 4.12 can be used to investigate the effects of different factors on the rate of photosynthesis.

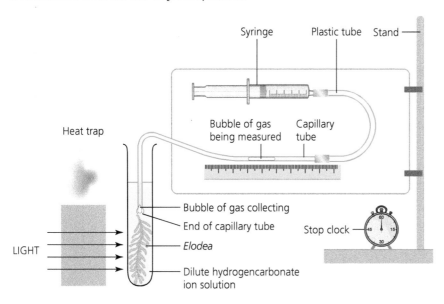

Figure 4.12 Measuring the rate of photosynthesis with a microburette. The heat trap on the left-hand side of the apparatus is a water-filled vessel; this ensures that only light affects the pondweed rather than heat from the light source.

Figure 4.12 shows how the rate of photosynthesis can be estimated using a microburette to measure the volume of oxygen given out at different light intensities.

■ ACTIVITY

Figure 4.13 shows a leaf attached to a plant throughout an experiment.

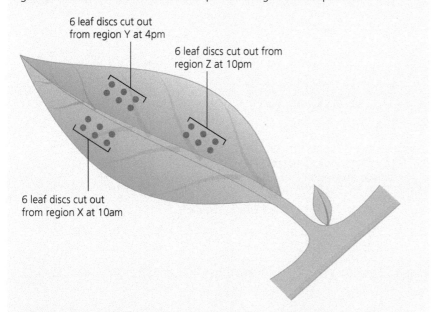

Figure 4.13 Leaf discs removed from a plant to investigate the effect of sunlight on photosynthesis

Expert tip

The apparatus used to measure the rate of production of oxygen from an aquatic plant is known as a photosynthometer. The bubbles produced are oxygen-enriched rather than pure oxygen.

Six discs were cut out from regions **X**, **Y** and **Z** of the leaf at 10 am, 4 pm and 10 pm respectively. The plant received some treatment before the discs were cut as did some of the discs before their total dry mass was measured. The table below shows the steps taken in the experiment.

Time	Treatment to plant before removing leaf discs	Treatment of leaf discs	Dry mass of leaf discs/mg
10 am	Exposed to sunlight	Exposed to steam	186
4 pm	Exposed to sunlight	None	209
10 pm	Kept in the dark for 6 hours	None	198

17 **Explain** the change in dry mass of leaf discs between 10 am and 4 pm and between 4 pm and 10 pm.

■ Investigating how carbon dioxide concentration affects the rate of photosynthesis

The same equipment is used as outlined above, although in this experiment the light intensity is kept constant. Different concentrations of $NaHCO_3$ (sodium hydrogencarbonate) are used to vary the concentration of carbon dioxide in the water (0 g L^{-1}, 20 g L^{-1}, 40 g L^{-1}, 60 g L^{-1}, and 80 g L^{-1}). The number of oxygen-enriched bubbles released per minute is counted for a fixed length of time. To do this, running totals must be calculated (see Expert tip below).

Expert tip

The raw data recorded in your data table will need processing and then transforming (graphing). This makes it easier to establish trends and relationships between variables.

Finding the total

This operation involves finding the total value of all the data or data entries in a particular column: first data entry + second data entry + … etc. For example, (15 + 23 + 37) = 75 (Table 4.2).

Time/s ± 0.5 s	Number of oxygen-enriched air bubbles produced
60.0	15
120.0	23
180.0	37
Total	75

Table 4.2 A data table showing a total

Maintaining a running total

This operation involves adding each new data value or data entry to the earlier ones in a column (see Table 4.3). The total of earlier data + next data entry = running total.

Time/s ± 0.5 s	Number of oxygen-enriched air bubbles produced	Total number of oxygen-enriched air bubbles produced
60.0	15	15
120.0	23	38
180.0	37	75

Table 4.3 A data table showing the running total

■ The Hill reaction

Aim

To isolate active chloroplasts and investigate the **Hill reaction**.

When chloroplasts absorb light, chlorophyll *a* pigment molecules use the harvested light energy to split water molecules (**photolysis**). The water molecules form hydroxide ions, hydrogen ions and electrons (e⁻) (Figure 4.14).

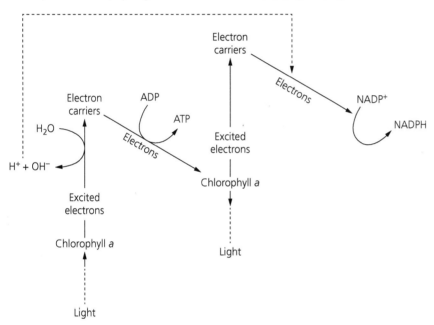

Figure 4.14 The photolytic breakdown of water (PSII and PSI not shown)

Chloroplasts can be isolated from green plant leaves and suspended in an ice-cold buffer solution of the same concentration as the cytosol (using an **isotonic** buffer). Suspended in such a buffer, it has been found that the chloroplasts are undamaged and function much as they do in the intact leaf. These isolated chloroplasts can be used to investigate the reactions of photosynthesis – for example, to show they evolve oxygen when illuminated (known as the Hill reaction – see Expert tip). This occurs provided the natural electron-acceptor enzymes and carrier molecules are present.

A hydrogen-acceptor dye that changes colour when it is reduced can be used to measure the rate of reaction. The dye known as DCPIP (2,6-dichloro-phenol-indophenol) is an example. DCPIP does no harm when added to chloroplasts in a suitable buffer solution, but changes from blue to colourless when reduced. The splitting of water molecules by light energy (photolysis) at photosystem II, is the source of hydrogen that turns DCPIP colourless.

The photolysis of water and the reduction of the dye are represented by the equation:

$$2DCPIP + 2H_2O \rightarrow 2DCPIPH_2 + O_2$$

The electrons involved in the reduction of DCPIP therefore come from the oxidation of water molecules by photosystem II:

$$2H_2O \rightarrow 4H^+ + 4e^- + O_2$$

Expert tip

Blue, oxidized DCPIP is reduced to the colourless form:

$$DCPIP \text{ (blue)} + 2H^+ + 2e^- \rightarrow DCPIPH_2 \text{ (colourless)}$$

The progress of the reaction can be measured by the change in absorbance at 600 nm of the DCPIP solution.

Expert tip

The 'dark' reactions on photosynthesis are better termed 'light-independent' reactions because they are still ongoing during exposure to light, with the light reactions being accelerated by light.

Equipment

- colorimeter with suitable cuvettes or tubes (see Chapter 2, pages 32–3 and page 35 for further information)
- lamps with 100 watt frosted bulbs
- domestic blender
- 10 ml volumetric pipettes
- centrifuge
- muslin cloth
- spinach leaves, or equivalent
- fine paintbrushes

Stock solutions

- Chloroplast isolation buffer: 50 mM tricine, 400 mM sorbitol, 10 mM NaCl, 2.5 mM $MgCl_2 \bullet 6H_2O$, 1.25 mM $MnCl_2$, 0.3 mM Na_2EDTA. Adjust pH to 7.8 using NaOH
- 0.2 mM DCPIP

Procedure

Extraction of chloroplasts

1 Remove the stalks and larger veins from fresh, chilled spinach leaves. Take approximately 50 g of leaves and cut the blade tissue into very small pieces.

2 Put the leaf tissue into a blender containing 120 cm^3 of ice-cold buffer solution.

3 Gently blend the tissue over a 10–15-second period, ensuring all the leaves become submerged in the buffer solution during the process. Then switch the blender to full speed for 5 seconds to form a homogeneous green liquid.

4 Filter the blended tissue through several layers of muslin, collecting the filtrate in a beaker standing on ice (to slow enzyme reactions).

5 Pour equal quantities of the filtrate into two centrifuge tubes and centrifuge at low speed to remove the heavier debris and the heaviest organelles. Then centrifuge for six minutes at the full speed of the centrifuge (3 000–5 000 rpm). The spinning action forces the chloroplasts down to the bottom of the tubes.

6 Discard the supernatant (the liquid lying above the solid chloroplast residue after centrifugation) in both centrifuge tubes.

7 Using a fine paintbrush, re-suspend each of the green chloroplast pellets at the base of the tubes in 5 cm^3 of ice-cold buffer solution. Combine these suspensions in one tube standing in the ice bath, and label it 'chloroplast suspensions'.

Investigation of the Hill reaction

8 Set up four reaction mixtures as shown in the table below, in colorimeter tubes (cuvettes), one at a time. *Only add the chloroplast suspension when you are ready to carry out the measurements.*

Tube	Buffer/cm³	DCPIP/cm³	Distilled water/cm³	Chloroplast suspension/cm³
1	1.5	0.2	0.0	2.3
2	1.5	0.2	0.0	2.3
3	1.5	0.2	2.3	0.0
4	1.5	0.0	0.2	2.3

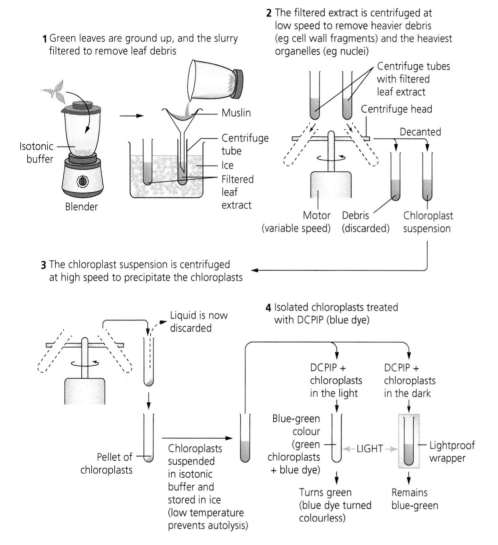

Figure 4.15 Demonstration of the Hill reaction in isolated chloroplasts

9 Carry out the following treatments to the tubes in turn. (*Note: tubes 2, 3 and 4 are all control experiments.*)

Tube 1

- Place the tube in the colorimeter with the correct filter in position and zero the 'absorbance scale'.

- Remove the tube, expose it to bright light, and return it after 30 seconds to the colorimeter.

- Read and record the absorbance value.

- Repeat the exposure/absorbance measurement cycle at 30-second intervals until no further change in the reading occurs.

Tube 2

- Place the tube in complete darkness (surround it with aluminium foil and place in a darkened container).

- Leave the tube in darkness for the same period of time that tube 1 took to show no further colour change.

- Place tube 2 into the colorimeter and record the absorbance.

Tubes 3 and 4

- Treat these tubes in the same way as tube 1, maintaining the cycles of exposure and measurement of absorbance for the same period of time as it took for tube 1 to show no further colour change.

10 Draw a graph of change in absorbance (*y*-axis) against time (*x*-axis) and record the results obtained from all four tubes. Label each curve with the tube number.

■ ACTIVITIES

In the experiment above (pages 68–9):

18 What change was observed in tube 1?

19 What was the purpose of tubes 2 and 3?

20 a What other organelle apart from chloroplasts might you expect to find in the chloroplast suspension?

   b What evidence would you have that these other organelles were not involved in the reduction of DCPIP?

21 Why was the isolation medium kept ice-cold?

22 Why was the isolation medium buffered?

23 In this experiment, what was:

   a the electron acceptor?   b the electron donor?

> **Expert tip**
>
> It might be necessary to alter the volumes of DCPIP so that the colorimeter will read on the given scale, and – depending on the activity of the chloroplast suspension – so that the colour change in tube 1 takes no more than 5–10 minutes.

Ideas for investigations

1 It is possible to determine the change in molarity, and therefore the number of moles of DCPIP reduced in a specific amount of time, by using a standard curve of absorbance at 600 nm against DCPIP concentration. 1 mole of DCPIP contains 6.0221×10^{23} (Avogadro's number) molecules of DCPIP, and so the number of molecules of DCPIP that have been reduced can be calculated.

Consequently, the following half equations can be used to calculate further quantities:

$$DCPIP + 2H^+ + 2e^- \rightarrow DCPIPH_2$$

$$2H_2O \rightarrow 4H^+ + 4e^- + O_2$$

Other quantities that can be estimated are:

- the number of electrons transferred to reduce DCPIP
- the amount of water oxidized to produce the electrons
- the volume of oxygen evolved.

2 The Hill reaction could be applied to several different situations. For example, the following could be examined: the effect of different types of weed-killer on isolated chloroplasts; the effect of different light intensities on chloroplasts taken from different types of plant, for example, spinach and barley; the effect of one plant species growing in different situations (shade versus bright sunshine); the difference between sun- and shade-loving plants.

- Other experiments could explore the role of uncoupling agents on the rate of decolourization of DCPIP. Uncouplers, such as ammonium chloride (NH_4Cl), act by separating the process of photophosphorylation from electron transport. When the reactions relating to electron transport and the creation of an H^+ gradient are uncoupled, electron transport proceeds at a faster rate, and so the reduction of DCPIP increases in rate.

- Temperature also affects the rate of the Hill reaction, and would be another factor to investigate.

3 Other ideas for plant physiology investigations:

- Seed **germination** can be studied, for example, the effect of temperature on the percentage success of germination.

- The effect of different fertilizers on plant growth, using cress, wheat, peas or beans. Fast-cycling brassicas have been specifically bred for use in schools and colleges and complete their life cycle within six weeks, making them ideal for study.

- During autumn, leaves change colour when different pigments become prominent as others degrade. Trees of the maple family undergo particularly notable changes in colour: these may be linked with higher levels of sugar in the tree. The colour changes in different tree species could be examined using chromatography, where different pigments in the leaves at different stages of colour change could be separated and identified. Species with more dramatic colour changes (such as maple) could be compared to those with less notable changes.

5 Ecology practical skills

Measuring abiotic components of the system

Abiotic factors include:

- **marine** – turbidity, salinity, pH, temperature, dissolved oxygen, wave action
- **freshwater** – turbidity, flow velocity, pH, temperature, dissolved oxygen
- **terrestrial** – temperature, light intensity, wind speed, particle size of soil, air content of soil, slope, soil moisture, drainage, mineral content.

Standardized methods are needed to compare **ecosystems**.

Abiotic factor	How is it measured?	Ecosystem	Evaluation
Wind speed	Anemometer	Terrestrial	Gusty conditions can lead to large variations in data
Temperature	Thermometer	Terrestrial, freshwater, marine	Problems in data reproducibility and accuracy if temperature is not taken from consistent depth
Light	Light meter	Terrestrial, freshwater, marine	Cloud cover changes light intensity, as does shading from plants or light-meter operator
Soil compaction	Penetrometer	Terrestrial	Readings must be taken in the same way, with the metal bolt (Figure 5.2) dropped from the same height
Flow velocity	Flow meter	Freshwater	Readings must be taken from same depth; water flow can vary due to rainfall/ice melt
Wave action	Dynamometer	Marine	Changes in wave strength during a day and over a monthly period affect results
Turbidity	Secchi disc	Marine	Reflections off water reduce visibility; measurements are subjective
Dissolved oxygen concentration (in ppm)	Dissolved oxygen meter	Freshwater	Possible contamination from air/oxygen bubbles in the **samples** when using dissolved oxygen meter
Soil moisture	Evaporate water; soil moisture probes	Terrestrial	If soil is too hot when evaporating water, organic content can also burn off

Table 5.1 The measurement of abiotic factors in ecosystems

Figure 5.1 Using an anemometer to measure wind speed in a shingle ridge succession

Figure 5.2 Using a simple penetrometer to measure soil compaction. A metal bolt with a pointed end is dropped within a guide sleeve – depth of penetration of soil indicates compaction.

Figure 5.3 Using a flow meter to measure water speed in a forest river

Abiotic factors can vary from day to day and season to season. Electronic data loggers overcome many of the limitations shown by abiotic measuring devices (page 106):

- They provide continuous data over a long period of time.

- They make data more representative of the area being sampled.

- More data can be collected, making results more reliable.

■ Soil moisture

Soils contain water and organic matter. You can estimate moisture levels by weighing samples before and after heating in an oven and calculating the mass of water evaporated. Repeated readings should be taken until no further mass loss is recorded – the final reading should be used. Loss of mass can be calculated as a percentage of the starting mass.

■ ACTIVITIES

1 **Identify** an abiotic factor found in a freshwater ecosystem. **Outline** how you would measure this factor.

2 **Identify** an abiotic factor found in a marine ecosystem. **Outline** how this factor would vary with depth.

3 **Identify** an abiotic factor found in a terrestrial ecosystem. **Evaluate** the technique used to measure this factor.

Measuring biotic components of the system

Standardized methods are needed to compare **biotic factors** of ecosystems with one another. Such studies also allow ecosystems to be monitored and evaluated over time, and for the effects of human disturbance to be understood.

▨ Keys

Your investigation might require you to identify and name different terrestrial or marine animals or plants. One approach is to use a book that has photographs of groups of related animals and plants and use that to identify them. Another approach is to use a biological key. Any keys or books that you plan to use in the identification of animals or plants should be outlined in your plan and later referenced in your bibliography (see pages 140–1).

Expert tip

When carrying out fieldwork you must follow the IB ethical practice guidelines and IB animal experimentation policy: that is, animals and the environment should not be harmed during your work.

▨ Naming of organisms

Binomial names should be used to describe organisms. Binomial means two names: the first name gives the genus and the second gives the species. For example, *Cardamine pratensis* is the binomial name for an English flower known as the 'cuckoo flower' or 'lady's smock'. The first name in the binomial name is the genus (group of closely related species) and the second name is the species name. The binomial name is written in Latin and italicized.

▨ Methods for estimating the abundance of organisms

The way in which the abundance of an organism is measured depends on whether it is **motile** or **non-motile**.

▨ Lincoln index

This technique uses the capture–mark–release–recapture method. It is used for estimating the population size of motile animals.

- Organisms are captured, marked, released and then recaptured.

- Marking varies according to the type of organism. For example, wing cases of insects can be marked with pen, snails with paint, and fur clippings can be used for mammals.

- Markings must be difficult to see – high visibility increases predation risk.

- The number of individuals of a species is recorded at each stage.

- The total population size is estimated using the following equation:

$$N = \frac{n_1 \times n_2}{m}$$

Where:

N = total population of animals in the study site

n_1 = number of animals captured (marked and released) on first day

n_2 = number of animals recaptured on second day

m = number of marked animals recaptured on second day

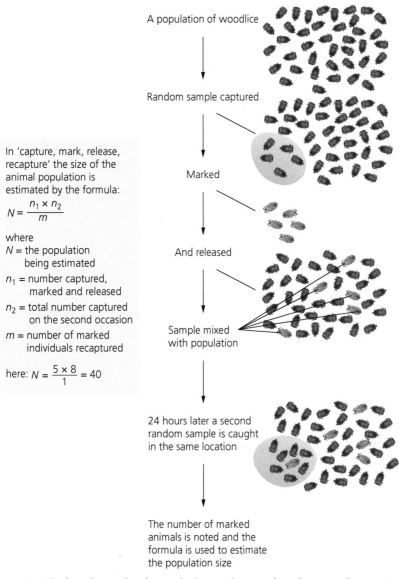

A population of woodlice

Random sample captured

Marked

And released

Sample mixed with population

24 hours later a second random sample is caught in the same location

The number of marked animals is noted and the formula is used to estimate the population size

In 'capture, mark, release, recapture' the size of the animal population is estimated by the formula:

$$N = \frac{n_1 \times n_2}{m}$$

where
N = the population being estimated
n_1 = number captured, marked and released
n_2 = total number captured on the second occasion
m = number of marked individuals recaptured

here: $N = \frac{5 \times 8}{1} = 40$

Figure 5.4 Estimating animal populations using mark, release and recapture

Figure 5.5 A quadrat being used to estimate the species richness of plants in a water meadow at Slapton Ley, Devon

■ **ACTIVITY**

4 **Describe** the way the capture–mark–release–recapture method is used to estimate population size using a named animal species.

▨ Quadrat methods

Quadrats are used for estimating the abundance of plants and non-motile animals.

■ A quadrat is a square frame that outlines a known area for the purpose of sampling (Figure 5.5).

■ Quadrats are placed according to random numbers, after the area has been divided into a grid of numbered sampling squares. The presence or absence in each quadrat of the species under investigation is then recorded.

■ **Percentage frequency** is the percentage of quadrats in an area in which at least one individual of the species is found. It is calculated by taking the number of occurrences and dividing by the number of possible occurrences; for example, if a plant occurs in 3 out of 100 squares in a grid quadrat, then the percentage frequency is 3%.

■ **Percentage cover** is the proportion of a quadrat covered by a species, measured as a percentage. It is worked out for each species present. Estimates can be made by dividing the quadrat into a 10 × 10 grid (100 squares), where each square is 1% of the total area covered.

■ **Population density** is the number of individuals of each species per unit area. It is calculated by dividing the number of organisms by the total area of the quadrats.

Key definition

Quadrat – a square frame which outlines a known area for the purpose of sampling.

Expert tip

Formulas do not need to be memorized for exams but should be applied to given data.

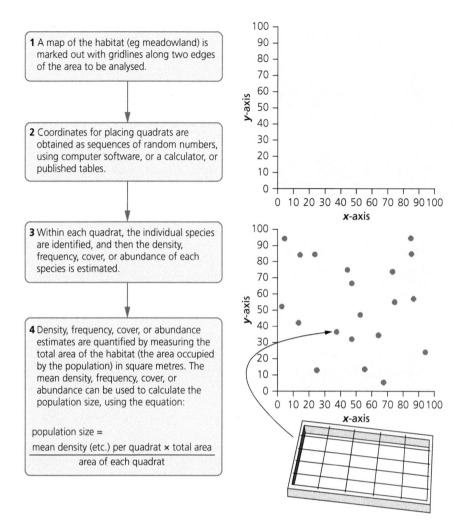

Figure 5.6 Random locating of quadrats

It is not possible to measure all organisms in an ecosystem, and so a sample must be taken. The **sample size** is the number of samples taken from a population.

The sampling system used depends on the areas being sampled:

- **Random sampling** is used if the same habitat is found throughout the area.

- **Stratified random sampling** is used in areas which contain two or more different habitat types. For example, if an area of woodland is being studied, there are likely to be different types of habitat within it: random sampling may miss one or more of these and so stratified sampling is used. This technique takes into account the proportional area of each habitat type within the woodland and samples each one accordingly. The technique can also be used to compare undisturbed and disturbed areas.

- **Systematic sampling** is used along a **transect** where there is an environmental gradient, such as the change from the edge of a woodland, adjoining open land, into interior forest where, for example, warmer and lighter conditions predominate at the edge of the forest and cooler, darker ones in the interior.

> **Key definitions**
>
> **Sample size** – the number of samples taken from a population.
>
> **Random sampling** – a method of choosing a sample from a population without any bias.
>
> **Transect** – arbitrary line through a habitat, selected to sample the community.

Expert tip

If you have an iPhone, you can convert it into a scientific calculator by turning it sideways (ie 'landscape') when in calculator mode. The scientific calculator has a random number generator function.

Expert tip

Random sampling ensures that every individual in the community has an equal chance of being selected and so a representative sample is assured.

Expert tip

Quadrats must be placed randomly to avoid sampling bias. Subjective choice of location for quadrats would lead to samples that are not representative of the area they are sampling. For example, areas that have a large number of species might be chosen at the expense of those with less species richness. Random allocation of sampling sites should always be used when a uniform habitat is being sampled.

Ideas for investigations

Sampling strategies may be used to measure biotic and abiotic factors and their change in space, along an environmental gradient (pages 78–80), or before and after a human impact for example.

Method for estimating the biomass of trophic levels

Biomass is calculated to indicate the total energy within a trophic level.

- Biomass is a measure of the organic content of organisms.

- Water is not an organic molecule, and its amount varies from organism to organism, so water is removed before biomass is measured. This is called **dry weight biomass**.

- One criticism of the method is that it involves killing living organisms (although not all the organisms in an area need to be sampled – see below).

- Problems exist with measuring the biomass of very large plants such as trees, and with roots and underground biomass.

Calculating dry weight biomass

To obtain quantitative samples, biological material is dried to constant mass:

- The sample is weighed in a container of known mass.

- The sample is put in a hot oven (80 °C).

- After a specific length of time the sample is reweighed.

- The sample is put back in the oven.

- This is repeated until the same mass is recorded from two successive readings.

- No further loss in mass indicates that water is no longer present.

Biomass is recorded per unit area (eg per metre squared) so that trophic levels can be compared. Not all organisms in an area need to be sampled:

- The mass of one organism, or the average mass of several organisms, is taken.

- This mass is multiplied by the total number of organisms to estimate total biomass.

- This is called an **extrapolation** technique.

Expert tip

Data from methods for estimating biomass can be used to construct ecological pyramids.

Key definition

Biomass – the mass of organic material in organisms or ecosystems, usually stated per unit area.

Expert tip

To estimate the biomass of a primary producer, all the vegetation, including roots, stems and leaves, is collected within a series of 1 m × 1 m quadrats. The dry weight method is carried out and average biomass calculated.

ACTIVITIES

5   **Describe** and **evaluate** methods for measuring **three** abiotic factors in a forest ecosystem.

6   **Explain** the difference between percentage frequency and percentage cover.

7   Which data are needed to estimate the size of an animal population? Write the equation needed to calculate population size.

8   **Explain** how biomass is calculated.

■ Diversity and the Simpson's diversity index

Species **diversity** refers to the number of species and their relative abundance. It can be calculated using a **diversity index** (plural indices).

Species diversity can be calculated using the **Simpson's diversity index**, using the equation:

$$D = \frac{N(N-1)}{\sum n(n-1)}$$

where:

- ■ D = Simpson's diversity index

- ■ N = total number of organisms of all species found

- ■ n = number of individuals of a particular species

- ■ $\sum$ = the sum of

Comparisons can be made between areas containing the same type of organism in the same ecosystem.

- ■ A high value of D suggests a stable and ancient site, where all species have similar abundance (or 'evenness').

- ■ A low value of D could suggest disturbance through, say, logging, pollution, recent colonization or agricultural management, where one species may dominate.

> ## Key definitions
>
> **Diversity** – a generic term for heterogeneity (that is, variation or variety). The scientific meaning of diversity becomes clear from the context in which it is used; it can refer to heterogeneity of species or habitat, or to genetic heterogeneity.
>
> **Diversity index** – a numerical measure of species diversity calculated by using both the number of species (species richness) and their relative abundance.

■ ACTIVITY

9 One habitat has a Simpson's diversity index of 1.83 and another has an index of 3.65. What do these values indicate about each habitat?

Worked example

The table below contains data from two different habitats. Total number of species (= 'species richness') and total number of individuals is the same in each case. Calculate the diversity of each habitat and comment on the differences between the habitats.

Species found	Number found in habitat X	Number found in habitat Y
A	10	3
B	10	5
C	10	2
D	10	36
E	10	4
Number of species =	5	5
Number of individuals =	50	50

The Simpson's index must be calculated for each habitat. This can be done using a table to calculate components of the index:

Species	Number (n) found in habitat X	$n(n-1)$	Number (n) found in habitat Y	$n(n-1)$
A	10	10(9) = **90**	3	3(2) = **6**
B	10	10(9) = **90**	5	5(4) = **20**
C	10	10(9) = **90**	2	2(1) = **2**
D	10	10(9) = **90**	36	36(35) = **1 260**
E	10	10(9) = **90**	4	4(3) = **12**
	$\Sigma n(n-1)$	**450**	$\Sigma n(n-1)$	**1 300**

Species diversity for each habitat:

Habitat X:

$$D = \frac{50(49)}{450} = \frac{2450}{450} = 5.44$$

Habitat Y:

$$D = \frac{50(49)}{1300} = \frac{2450}{1300} = 1.88$$

What do these values say about each habitat?

- Greater 'evenness' between species in habitat X.
- Less competition due to non-overlapping niches in habitat X.
- One species does not dominate in X, reflecting greater habitat complexity/more niches.
- Habitat Y is less complex with fewer/overlapping niches, where one species can dominate, leading to lower diversity.

■ **ACTIVITY**

10 Consider a man-made park in your city or town. **Suggest** what variables might be important in determining the relative species richness of trees and other plants.

Measuring changes along an environmental gradient

Ecological gradients are found where two ecosystems meet or where an ecosystem ends. Abiotic and **biotic** factors change along the same ecological gradient.

Transects are used to measure changes along the gradient, to ensure that all parts of the gradient are measured (Figure 5.7):

- The whole transect can be sampled – a **continuous** transect – or samples can be taken at points of equal distance along the transect – an **interrupted** transect.

- A **line transect** is the simplest transect, where a tape measure is laid out in the direction of the gradient. All organisms touching the tape are recorded.

- A **belt transect** allows more samples to be taken – a band usually between 0.5 m and 1.0 m is sampled along the gradient.

Figure 5.8 A point quadrat being used to measure plant species richness along a transect

Figure 5.7 Sampling along an environmental gradient. A tape measure is laid out at 90° to the sea and abiotic and biotic factors measured at regular intervals along it. The photo shows a site near the end of the succession which starts on a shingle ridge and ends in shrub communities.

Quadrats can be used to sample at regular intervals along a transect (see pages 74–6):

■ **Frame quadrats** are empty frames of known area (eg 1 m²).

■ **Grid quadrats** are frames divided into 100 small squares.

■ **Point quadrats** are made from a frame with 10 holes, inserted into the ground by a leg (see Figure 5.8). They are used for sampling vegetation that grows in layers. A pin is dropped through each hole in turn and the species touched are recorded. The total number of pins touching each species is converted to percentage frequency data (ie if a species is touched by 7 out of the 10 pins it has 70% frequency).

■ **ACTIVITIES**

11 **Distinguish** between using a transect and a quadrat in collecting ecological field data.

12 An investigation was carried out on the effect of pesticide treatment on earthworm populations. The results are shown in the table.

Quadrats on soil treated with pesticide		Quadrats on untreated soil	
Worms per quadrat	Frequency	Worms per quadrat	Frequency
0	0	7	0
1	1	8	2
2	3	9	3
3	4	10	6
4	6	11	12
5	10	12	9
6	9	13	6
7	5	14	4
8	4	15	2
9	3	16	1
10	0	17	0

(a) Plot a histogram of frequency against numbers of earthworms per quadrat.

(b) **Describe** the effect of pesticides on earthworm populations and **suggest** a reason for the observed results.

Zonation can be measured by recording biotic and abiotic factors at fixed heights along a transect:

■ A cross staff is used to move a set distance (eg 0.6 m) vertically up the transect (Figure 5.9).

■ The staff is set vertically and a point measured horizontally from an eye-sight 0.6 m from the base of the staff.

■ Biotic and abiotic factors are measured at each height interval.

Figure 5.9 A cross staff being used to relocate quadrats at regular height intervals along a rocky shore. This allows zonation on the shore to be studied.

Expert tip

Studying both biotic and abiotic factors allows research questions such as *how do abiotic factors affect the distribution of organisms in ecosystems?* Different species can be expected to be found at different locations along the gradient, as they will be adapted to different conditions.

■ **ACTIVITIES**

13 What is meant by the term *ecological gradient*?

14 **Outline** how you would measure changes in abiotic factors along an environmental gradient.

15 **Describe three** different methods for recording biotic factors along a belt transect.

16 **Describe** how you would collect data to show zonation in an ecosystem.

Common mistake

Do not confuse the terms biotic and abiotic – biotic refers to the living parts of an ecosystem and abiotic to the non-living parts.

Practical 5: Setting up sealed mesocosms to try to establish sustainability

A **mesocosm** can be set up to investigate eutrophication, so avoiding the destruction of a natural ecosystem, and allowing one variable to be altered and the rest controlled.

Mesocosms are enclosed experimental areas that are set up to explore ecological relationships. Because they are contained experimental areas, they can be closely controlled and variables can be monitored. It can be difficult to study natural ecosystems because there are so many variables that cannot be controlled – mesocosms enable all variables other than the independent and dependent variables to be kept constant, so as to ensure a fair test.

Key definition

Mesocosm – enclosed experimental area that is set up to explore ecological relationships. Because it is a contained experimental area it can be closely controlled and variables monitored.

Mesocosms can be set up in open tanks, but sealed glass vessels are preferable because entry and exit of matter can be prevented but light can enter and heat can leave. Aquatic systems are likely to be more successful than terrestrial ones.

The sustainability of an ecosystem can change when an external 'disturbing' factor that disrupts the natural balance is applied. An investigation into the effects of a disturbing factor can be attempted in a natural habitat or in an experimental, enclosed system. A mesocosm can be used as an enclosed system to investigate the effect of altering one variable on the stability of an ecosystem, and to establish whether the changes are sustainable or not. Both approaches have advantages and drawbacks, as detailed in Table 5.2.

	A natural ecosystem, eg an entire pond or lake	A small-scale laboratory model aquatic system (a mesocosm)
Advantages	Realistic – actual environmental conditions experienced	Able to control variables; opportunity to measure degree of stability/change in a community, and to investigate the precise impact of a disturbing factor
Disadvantages	Variable conditions – minimum or non-existent control over 'controlled variables'	Unrealistic – possibility of disputed relevance/applicability to natural ecosystem

Table 5.2 Comparison between ecosystem and mesocosm studies

Setting up a mesocosm

You can set up a mesocosm to investigate the effect of changing one variable on the sustainability of the system. For example, in an aquatic system you could set up one mesocosm with fish and another without fish to investigate the effect of fish on the aquatic ecosystem.

When setting up a mesocosm you need to consider various factors.

- What variables will you control (keep the same) and why? What effect might these variables have on the system if they are changed (ie why do you need to control them)?

- For terrestrial mesocosms, large glass jars can be used, although plastic containers can be just as effective. Which will you use, and why? Should the sides of the container be transparent or opaque?

- Which groups of organism will you need to include in the mini-ecosystem? Think of the organisms that would be present in a natural ecosystem (eg autotrophs, consumers and decomposers).

- Because the mesocosm will be sealed, you need to consider how the organisms will obtain fresh sources of oxygen. Photosynthetic organisms will ensure a supply of oxygen – how will you ensure these are kept alive?

- Because the mesocosm is a closed system, there is a danger that the organisms in it will suffer from lack of food, competition, excess heat and so on. How will you ensure the well-being of the organisms in your mesocosm?

Case study: An investigation of eutrophication

Lakes and ponds where there is an excess of nutrients can undergo a process called eutrophication. Excess nutrients might come from, for example, fertilizer run-off from surrounding land. The excess nitrates and phosphates provide high levels of nutrition for algae, which undergo rapid population growth (an algal bloom occurs). The algae block light to underwater plants which die, providing detritus for bacteria to absorb and use for respiration. These bacteria undergo rapid growth and remove oxygen from the water, causing fish and other aquatic animals to die. A few organisms can survive in these conditions and prosper, but the death of many aquatic organisms results.

Expert tip

In water enriched with inorganic ions, the increased in concentration of ammonium, nitrate and phosphate ions increases algal and plant growth.

A mesocosm can be set up to investigate eutrophication, thereby avoiding the destruction of a natural ecosystem, and allowing one variable to be altered and the rest controlled (Figure 5.10).

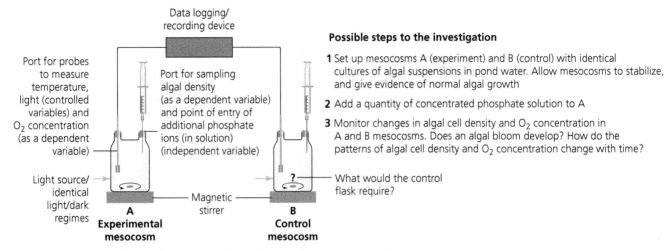

Possible steps to the investigation

1 Set up mesocosms A (experiment) and B (control) with identical cultures of algal suspensions in pond water. Allow mesocosms to stabilize, and give evidence of normal algal growth

2 Add a quantity of concentrated phosphate solution to A

3 Monitor changes in algal cell density and O_2 concentration in A and B mesocosms. Does an algal bloom develop? How do the patterns of algal cell density and O_2 concentration change with time?

Figure 5.10 An experimental mesocosm investigating the effect of nutrient enrichment

In this experiment:

■ The independent variable is the volume of phosphate ions of fixed concentration added to mesocosm A.

■ The dependent variable is either dissolved oxygen concentration, or algal density.

■ Controlled variables are temperature, light intensity, degree of stirring of solution.

Further reading and ideas

■ Mini-ecosystem ideas: http://www.biologycorner.com/worksheets/ecosystem. html#.U0rw2VVdXww

■ Create an ecosystem in a bottle: http://www2.nau.edu/lrm22/ lessons/bottle_biology/

■ Aquatic mesocosms: http://plankt.oxfordjournals.org/content/23/10/1081.full

■ Aquatic (ocean acidification): https://www.sciencedaily.com/ releases/2013/09/130913085756.htm

■ Freshwater, and why use mesocosms: http://www.whoi.edu/cms/files/spivak_et_ al_2011_fw_biol_80204.pdf

■ Biosphere 2: http://archive.bio.ed.ac.uk/jdeacon/biosphere/mesocos.htm

Planning ecological investigations

Ecology is the study of the interaction between organisms and their environment. In any given ecosystem, there are numerous complex, interconnected relationships. It is important in an ecological investigation to have a clear idea about the organisms that will be the focus of the investigation, and the relationship that is to be studied. You have to be careful that your investigation does not get too wide in scope, and has clearly defined parameters. Initial ideas about possible investigations can be gained from visiting a local nature reserve, woodland or other ecosystem. You might notice, for example, that certain species are only found in particular areas. You could test various abiotic factors, such as pH, light intensity, soil moisture and so on, to explore the reason for such patchy distribution, as well as sampling other species to see if there are any biotic interactions. Based on initial studies, you then need to choose one specific independent variable to investigate, and one dependent variable, such as the effect of soil moisture on the distribution of one species. Other confounding variables, which might have an effect on the dependent variable but which cannot be controlled, must be monitored.

Ecological investigations can draw on the range of techniques and skills you have learned in this chapter, such as the use of:

■ quadrats

■ transects

■ keys to identify organisms

■ abiotic sampling techniques

■ biotic sampling techniques

■ laboratory skills, eg measuring and calculating soil moisture

■ diversity indices.

The sampling methodology will be central to your investigation. Will you carry out random sampling (appropriate if your study area is homogeneous) or sample along a line transect (suitable where there is an environmental gradient and a change in species composition)?

Statistical tests are usually used in ecology to investigate whether observed results are significant or not (ie whether they are due to chance, or show that there is a real link between independent and dependent variables). It is important to select the appropriate test for your investigation. The use of statistical tests in ecological investigations is covered in the next chapter (Chapter 6, pages 89–100).

Ideas for investigations

Possible ecological investigations are almost limitless, but here are a few suggestions:

● Determine whether there is a correlation between the distribution of a plant species and an abiotic factor, such as light intensity.

● Explore whether there is an association between plant species, or between a plant species and an animal. For example, thyme (*Thymus serpyllum*) is often observed growing on anthills.

● Rocky shores provide many opportunities for investigations:

  ● Whether there is a correlation between height and width of molluscs, eg limpets (*Patella vulgata*), and exposure to wave action. Sheltered and exposed shoreline could be studied.

  ● Whether location of different species of seaweed on the shore is determined by their resistance to desiccation, eg seaweeds higher on the shore, which spend a greater proportion of their time not submerged by the sea, might be more tolerant of desiccation than ones lower on the shore. Seaweeds could be dried out for specific periods of time and percentage loss of water recorded for each species.

  ● Crabs make good subjects for ecological investigations, for example, fiddler crabs (*Uca* spp.), ghost crabs (*Ocypode* spp.) and soldier crabs (*Mictyris* spp.). They play an important role in the nutrient cycling and energy flow of coastal ecosystems, and can be used as indicators of ecosystem disturbances. See: http://ecolabnie.wixsite.com/shirley/research.

● Acid rain has an adverse effect on aquatic plants. The effect of acidity on plant growth could be studied.

● Lichens (a symbiotic association between fungal and algal species) are sensitive to air pollution, specifically sulfur dioxide. The abundance and distribution of lichens in industrial and rural areas could be studied and compared.

● Plant leaves often accumulate unpleasant tasting chemicals called tannins. These chemicals are used by plants to discourage insects from eating them. In general, older leaves have more tannins in them. The concentration of tannins can be measured by boiling leaves in water and adding a solution of iron(III) chloride to the boiled sample. The intensity of the resulting black colour is an indication of the concentration of tannin in the sample. The colorimetry methods described on pages 32–3 could be used to quantitatively compare samples.

6 Statistical analysis for ecological studies

Introduction

Data in ecological studies are usually complex and trends unclear. Statistical techniques allow data to be analysed and underlying patterns revealed.

Tabulating raw and processed data, drawing line graphs, calculating values such as mean, median, mode and standard deviation are all referred to as **descriptive statistics** and are useful for summarizing data and for assessing the variation in samples and sets of replicates. Their use helps to identify trends and draw conclusions from the data.

Statistical tests, sometimes called **inferential statistics**, involve complex calculations that allow differences to be compared between experimental treatments or samples to see if they are likely to have occurred as a result of variation in the data or if they are a treatment effect.

If the statistics show a treatment effect then the results are said to be statistically significant. A statistically significant result is one where there is a less than 5 % probability that it has occurred by chance alone. Confidence intervals or error bars indicate the variability of data around a mean.

■ Normal distribution

Data obtained from biological experiments may show a 'normal distribution': this means that when the frequency of particular classes of measurement is plotted against the classes of measurement, a symmetrical bell-shaped curve is obtained (Figure 6.1). Explanations of mode, median and mean are found below.

Normal distribution curve

Most biological data show variability, but with values grouped symmetrically around a central value.

Here the mode, median and mean coincide.

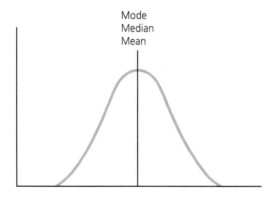

Skewed distribution

Values reduce in frequency more rapidly on one side of the most frequently obtained value than on the other.

Here the difference between the mean and mode is a measurement of 'skewness' of the data.

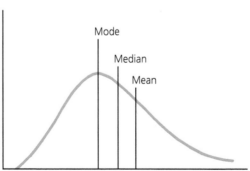

Figure 6.1 Frequency distributions of normal and skewed (non-normal) data

An example of a normal distribution is the number of humans at any particular height plotted against their different height classes, arranged in ascending order (Figure 6.2).

For the purpose of the graph, the heights are collected into arbitrary groups, each with a height range of 2 cm.

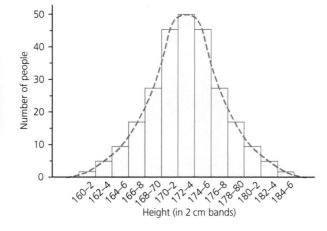

Figure 6.2 Normal distribution in human height data

Descriptive statistics: mode, median and mean

The normal distribution shows symmetrical distribution of data around the central tendency (a central value for a probability distribution). There are three different ways of calculating the central tendency:

- The **mode**: the most frequent value in a set of values.

- The **median**: the middle value in a set of values arranged in ascending order. If there is an even number of items in a data set then the median is found by taking the average (mean) of the two middle numbers.

- The average or arithmetic **mean**: calculated by dividing the sum of the individual values by the number of values obtained. The formula for the arithmetic mean is:

$$\bar{x} = \frac{\sum x}{n}$$

- where $\bar{x}$ = arithmetic mean

  $\sum x$ = the sum of all the measurements

  n = the total number of measurements.

Expert tip

The mode is not typically used as a measure of central tendency in biology, but it can be useful in describing a bimodal distribution, which has two peaks or modes. This type of distribution can be caused by disruptive selection.

Standard deviation

Standard deviation from the mean measures how spread out data are from the central tendency. It is a measure of the variation from the mean of a set of values.

- A small standard deviation indicates that the data are clustered closely around the mean value.

- A large standard deviation indicates a wider spread around the mean.

Once obtained, the value may be applied to the normal distribution curve (Figure 6.3). Note that 68 % of the data occurs within one standard deviation of the mean and more than 95 % of the data occurs within two standard deviations of the mean. So, a small standard deviation indicates that the observations (the values) differ very little from the mean. Standard deviation is also a measure of reliability: a small standard deviation indicates reliability.

Key definition

Standard deviation – the spread of a set of data from the mean of the sample; it is a measure of the variability of a population from a sample. A small standard deviation indicates that the data are more reliable.

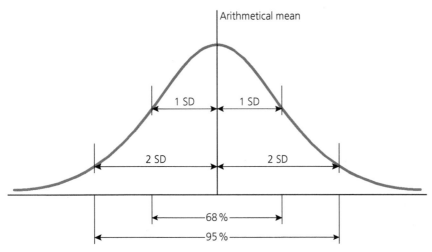

Figure 6.3 The normal distribution and its standard deviation (SD)

The standard deviation (*s*) is calculated in five steps.

1 First, calculate the mean, $\bar{x}$.

2 Calculate the deviation of each value from the mean: $x - \bar{x}$.

3 Square each deviation: $(x - \bar{x})^2$.

4 Add the squared deviations: $\Sigma(x - \bar{x})^2$.

5 Finally, divide by the number of values (*n*).

Worked example

Number of fruits	Frequency	
	R. acris	*R. repens*
15	0	0
16	0	1
17	0	1
18	1	1
19	2	2
20	1	4
21	4	4
22	8	8
23	7	7
24	9	9
25	10	10
26	16	16
27	9	9
28	10	10
29	4	4
30	5	5
31	3	3
32	1	1
33	1	1
34	2	2
35	1	1
36	1	1
37	0	0
38	0	0

An ecologist investigated the reproductive capacity of two species of a common grassland flower *Ranunculus acris* and *R. repens*. The latter species spreads vegetatively via strong and persistent underground stems. Would the use of vegetative reproduction result in fewer fruits from sexual reproduction compared to the other species that reproduces only by fruit production? Using comparable sized plants growing under similar conditions in the same soil, the ecologist counted and recorded the numbers of fruits formed in 100 flowers of each species. The results are given in the table and calculations of the standard deviations are shown in Figure 6.4.

Number of fruits	Frequency	
	R. acris	*R. repens*
39	2	0
40	2	0
41	0	0

Fruit production in *Ranunculus acris*

Values obtained in ascending order, x	Frequency f	fx	Deviation of x from the mean $(x - \bar{x})[= d]$	d^2	fd^2
18	1	18	−12	144	144
19	1	19	−11	121	121
20	1	20	−10	100	100
21	1	21	−9	81	81
22	1	22	−8	64	64
23	3	69	−7	49	147
24	4	96	−6	36	144
25	4	100	−5	25	100
26	5	130	−4	16	80
27	5	135	−3	9	45
28	6	168	−2	4	24
29	8	232	−1	1	8
30	14	420	0	0	0
31	12	372	1	1	12
32	10	320	2	4	40
33	7	231	3	9	63
34	3	102	4	16	48
35	2	70	5	25	50
36	3	108	6	36	108
37	2	74	7	49	98
38	3	114	8	64	192
39	2	78	9	81	162
40	2	80	10	100	200
$\Sigma f = 100$		$\Sigma fx = 2999$			$\Sigma fd^2 = 2031$

Mean of data $= \dfrac{\Sigma fx}{\Sigma f} = \dfrac{2999}{100} = 29.99$

$SD = \sqrt{\left(\dfrac{\Sigma fd^2}{\Sigma f - 1}\right)} = \sqrt{\dfrac{2031}{99}} = \sqrt{20.51} = 4.53$

Thus the mean of the sample *Ranunculus acris* = 29.99, and the SD = 4.53.

Fruit production in *Ranunculus repens*

Values obtained in ascending order x	Frequency f	fx	Deviation of x from the mean $(x - \bar{x})[= d]$	d^2	fd^2
16	1	16	−9	81	81
17	1	17	−8	64	64
18	1	36	−7	49	49
19	2	76	−6	36	72
20	4	80	−5	25	100
21	4	168	−4	16	64
22	8	154	−3	9	72
23	7	207	−2	4	28
24	9	240	−1	1	9
25	10	400	0	0	0
26	16	234	1	1	16
27	9	270	2	4	36
28	10	112	3	9	90
29	4	145	4	16	64
30	5	90	5	25	125
31	3	31	6	36	108
32	1	32	7	49	49
33	1	66	8	64	64
34	2	34	9	81	162
35	1	35	10	100	100
36	1	36	11	121	121
$\Sigma f = 100$		$\Sigma fx = 2479$			$\Sigma fd^2 = 1474$

Mean of data $= \dfrac{\Sigma fx}{\Sigma f} = \dfrac{2479}{100} = 24.79$

$SD = \sqrt{\left(\dfrac{\Sigma fd^2}{\Sigma f - 1}\right)} = \sqrt{\dfrac{1474}{99}} = \sqrt{14.89} = 3.86$

Thus the mean of the sample *Ranunculus repens* = 24.79, and the SD = 3.86.

Figure 6.4 Calculating the means and standard deviations of the *Ranunculus* reproductive capacity data

◼ Standard error

The **standard error** (S_M) represents how well the sample mean approximates to the population mean. The larger the sample, the smaller the standard error, and the closer the sample mean approximates to the population mean. The standard error is obtained by dividing the standard deviation, s, by the square root of n, the sample size.

$$\text{standard error, } S_M = \frac{\text{standard deviation, } s}{\sqrt{n}}$$

When graphs are presented showing mean values, error bars are added to each plotted value to demonstrate the deviation of the sample from the true population mean. Error bars ($\pm 1S_M$) extend above and below the points plotted on a graph to show this variability (Figure 6.5). Non-overlapping error bars demonstrate that the difference between mean values is significant (as shown in Figure 6.5), whereas overlapping error bars suggest a non-significant difference.

Calculating the standard errors (S_M):

	Standard deviation, s	$\frac{s}{\sqrt{n}}$	Standard error, S_M
R. acris	4.53	$\frac{4.53}{10}$	0.453
R. repens	3.86	$\frac{3.86}{10}$	0.386

Common mistake

A common misconception is that standard deviation decreases with increasing sample size. Standard deviation can either increase or decrease as sample size increases; it depends on the measurements in the sample. If there is a lot of variation in a population, the standard deviation will be large.

Key definition

Standard error – an estimate of the reliability of the mean of a population sample. A small standard error indicates that the mean value is close to the actual mean of the population.

Adding standard errors to a display of the means

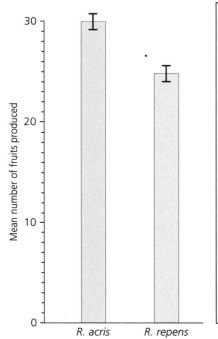

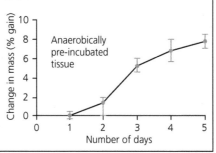

An example of the addition of error bars to a graph:

In an experiment on the effect of anaerobic pre-treatment of tissue discs on their subsequent change in mass, samples of 10 thin discs of plant tissue were used. (Thin cut discs allow all cells in a sample to receive more or less identical conditions.) The results of this enquiry indicated that anaerobic pre-treatment of the discs leads to a gain in mass. Error bars have been added to the curve in this graph.

Figure 6.5 Standard error and error bars

Inferential statistics: Statistical tests

All ecological statistical techniques involve hypothesis-testing. They test a statement called the **null hypothesis**. Statistical analyses test whether data match the null hypothesis or significantly vary from it. Where data are being compared, the null hypothesis states that 'there is no difference between the sets of data', and when an association is being investigated it states that 'there is no association'. The **hypothesis** is the 'opposite' of the null hypothesis, that is, that there *is* a difference or association shown by the data.

Because of the complexity of the data, ecologists can never be 100 % certain whether their results are true or not. Statistical tests allow them to be 95 % certain that any associations or **correlations** found in the data are real and not due to chance (that is, there is a 5 % chance that they *are* due to chance). The outcome of a statistical test is therefore a *probability* that the null hypothesis is true. A probability (known as the *p* **value**) varies from 0 (impossible) to 1 (certain). Since the *p* values are small, they are given as a percentage (0 to 100 %) to avoid possible confusion with small numbers. The lower the probability, the less likely it is that the null hypothesis is true.

One statistical test, known as the '*t*-test' (see Worked example, below), compares the means of data to test for significant differences between samples. For the *t*-test, the null hypothesis is 'There is no difference between the means of two samples'. By convention in biology, the **5 % significance level** (also known as the **95 % confidence level**) is used. In other words, if the probability is greater than 0.05 (5 %) then the null hypothesis is accepted. However, if the probability is 0.05 or less ($p < 0.05$), then the null hypothesis is rejected. This implies the event is predicted to happen by chance less than once in 20 times. So the difference is judged to be significant.

Charts of statistical probability are used to determine whether a result is significant or not (Figure 6.6). These present long series of data, at different levels of probability depending on sample size. These **critical values** are compared with the result of the statistical test – in general, if the result is greater than the critical value (the Mann-Whitney U test is an exception) then the null hypothesis is rejected and the hypothesis is accepted. Sample size is represented in a table of critical values as 'degrees of freedom'. In the *t*-test, for example, the number of degrees of freedom is determined by the number of samples in each data set, using the formula:

degrees of freedom = $(n_1 - 1) + (n_2 - 1)$

where n_1 = the number of samples in the first sample and n_2 = the number of samples in the second sample.

Each statistical test has a different method for determining degrees of freedom, although all are based on the number of independent observations in a set of data.

> ## Key definitions
>
> **Null hypothesis** – there is **no** statistically significant difference between two variables.
>
> **Hypothesis** – there **is** a statistically significant difference between two variables.
>
> **Correlation** – when one variable changes with another variable, so there is a relation between them. The strength of a correlation can be measured using a correlation coefficient. A correlation need not be a causal relation.

Degrees of freedom (df)	p values			
	0.10	**0.05**	**0.01**	**0.001**
1	6.31	12.71	63.66	636.60
2	2.92	4.30	9.92	31.60
3	2.35	3.18	5.84	12.92
4	2.13	2.78	4.60	8.61
5	2.02	2.57	4.03	6.87
6	1.94	2.45	3.71	5.96
7	1.89	2.36	3.50	5.41
8	1.86	2.31	3.36	5.04
9	1.83	2.26	3.25	4.78
10	1.81	2.23	3.17	4.59
12	1.78	2.18	3.05	4.32
14	1.76	2.15	2.98	4.14
16	1.75	2.12	2.92	4.02
18	1.73	2.10	2.88	3.92
20	1.72	2.09	2.85	3.85
22	1.72	2.08	2.82	3.79
24	1.71	2.06	2.80	3.74
26	1.71	2.06	2.78	3.71
28	1.70	2.05	2.76	3.67
30	1.70	2.04	2.75	3.65
40	1.68	2.02	2.70	3.55
60	1.67	2.00	2.66	3.46
120	1.66	1.98	2.62	3.37
∞	1.64	1.96	2.58	3.29

Decreasing value of p ⟶

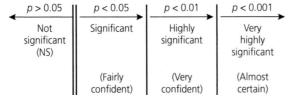

$p > 0.05$	$p < 0.05$	$p < 0.01$	$p < 0.001$
Not significant (NS)	Significant	Highly significant	Very highly significant
	(Fairly confident)	(Very confident)	(Almost certain)

Figure 6.6 Critical values for the *t*-test

■ Carrying out statistical tests

Statistical tests using standard deviation or standard error typically compare large, randomly selected representative samples of normally distributed data. In practice it is often the case that data can only be obtained from relatively small samples. Different statistical tests are used for different types of data (see Figure 6.7, page 93 and Table 6.1, page 92). The *t*-test, for example, may be applied to sample sizes of more than 15 and less than 30 taken from normally distributed data, and provides a way of measuring the overlap between two sets of data. A large value of *t* indicates little overlap and makes it highly likely there is a significant difference between the two data sets.

Even though the method for each test is different, the protocol for carrying out statistical tests is the same for each test. The following worked example shows how a statistical test should be presented.

Key definition

Statistical significance – a calculated value that is used to establish the probability that an observed trend or difference represents a true difference that is not due to chance alone.

Worked example

Applying the *t*-test

An ecologist was investigating woodland microhabitats, contrasting the communities in a shaded position with those in full sunlight. One of the plants was ivy (*Hedera helix*), but relatively few ivy plants were found in the locations under investigation. The following research question was developed: were the leaves in the shade actually larger than those in the sunlight?

Leaf widths were measured but, because the size of the leaves varied with their position on the plant, only the fourth leaf from each stem tip was measured. The results from the plants available are shown in the table.

Size-class/mm	Widths of leaves from plants in the sun (a)	Widths of leaves from plants in the shade (b)
20–24	24	
25–29	26, 26	26
30–34	30, 31, 31, 32, 32, 33	33, 34
35–39	37, 38	35, 35, 36, 36, 36, 37
40–44	43	41, 42
45–49		45

1   The null hypothesis assumes the difference under investigation has arisen by chance; that is, there is no difference in width between leaves from plants growing in sun and shade. The role of this statistical test is to determine whether to accept or reject the null hypothesis. If it is rejected in this case, we can have confidence that the difference in the leaf sizes of the two samples is statistically significant.

2   Check that the data are approximately normally distributed. This is done by arranging the data for the two samples and plotting a histogram, as shown in Figure 6.2.

3   The formula for the *t*-test for unpaired samples (see Expert tip below) (data sets 'a' versus 'b') is:

$$t = \frac{\bar{x}_a - \bar{x}_b}{\sqrt{\left(\dfrac{s_a^2}{n_a} + \dfrac{s_b^2}{n_b}\right)}}$$

*   where:
    $\bar{x}_a$ = the mean of data set a
    $\bar{x}_b$ = the mean of data set b
    s_a^2 = the standard deviation of data set a, squared
    s_b^2 = the standard deviation of data set b, squared
    n_a = the number of data items in set a
    n_b = the number of data items in set b

4   Once a value of *t* has been calculated (the value of *t* in this case is 2.10) a table of critical values for the *t*-test needs to be consulted. The degrees of freedom (df) for the two samples are determined using the formula: $(n_a - 1) + (n_b - 1)$ (see page 89). In this case, there are 11 + 11 = 22 degrees of freedom.

5   A table of critical values for the *t*-test is given in Figure 6.6. The column of significance levels (*p*) at the 0.05 level is read until the line corresponding with 22 degrees of freedom is reached. In this case, *p* = 2.08.

6   The calculated value of *t* (2.10) exceeds this critical value (2.08) at the 0.05 level of significance. This indicates that there is a lower than 0.05 (5 %) probability that the difference between the two means is solely

due to chance. Therefore, the null hypothesis can be rejected, and it can be concluded that **the difference between the two samples is significant**. (*For the experimenter, the significance of this statistic suggests there is a reason for the difference in the means. This can be further investigated and perhaps a fresh hypothesis proposed.*)

Expert tip

Scientific experiments often consist of comparing two or more sets of data. These data are described as **unpaired** or independent when the sets of data arise from separate individuals, and **paired** when they arise from the same individual at different points in time.

The unpaired *t*-test, for example, tests the null hypothesis that the population means related to two independent, random samples from an approximately normal distribution are equal.

Expert tip

A statistically significant result is one that is unlikely to be due to chance alone. Confidence intervals or error bars are used to indicate the variability of data around a mean. If the treatment average differs from the control average sufficiently for their confidence intervals not to overlap then the data can be said to be different.

Selecting the correct statistical test

Different statistical tests are used to analyse different types of data. Table 6.1 outlines the different tests available and how they should be used. Figure 6.7 can be used to select the correct statistical technique.

Statistical test	When to use it	Criteria for using test	How to interpret the value that is calculated
t-test	You want to know if two sets of continuous, normally distributed data are significantly different from one another.	You have two sets of continuous data with >15 but <30 readings for each set of data. Both sets of data are from populations that have normal (Gaussian) distributions with similar standard deviations.	Use a *t*-test table to look up the value of *t*. If this is greater than the *t* value for a probability of 0.05 then you can state that the two populations are significantly different.
Mann-Whitney *U*	You want to know if two sets of continuous, non-normally distributed data are significantly different from one another.	You have two sets of continuous data, with 6–20 pairs of data. The sets of data do **not** have normal distributions (that is, they are skewed).	*U* is calculated for each set of data. The lower value of *U* is used as the test statistic. Use a Mann-Whitney *U* table to look up the value of *U*. If this is **less than** or equal to the *U* value for a probability of 0.05 then you can state that the two populations are significantly different.
Chi-squared test	You want to know if your observed results differ significantly from your expected results.	You have two (or more) sets of nominal, categorical data.	Use a χ^2 table to look up the value of χ^2. If this value is greater than the χ^2 value for a probability of 0.05, then you can state that your observed results differ significantly from your expected results.
Pearson correlation coefficient	You want to know if there is a linear correlation between two paired sets of data.	You have two sets of interval data. At least 10 pairs of data, ideally more than 25. A scatter graph (see page 94) suggests there might be a linear relationship and both sets of data have an approximately normal distribution.	A value close to +1 indicates a positive linear correlation. A value close to −1 indicates a negative/inverse linear correlation. A value close to 0 indicates no correlation.
Spearman's rank	You want to know if there is a correlation (not necessarily linear) between two paired sets of data.	You have quantitative data that can be ranked. The samples for each set of data were made randomly. You have at least 10 pairs of data but ideally between 10 and 30. A scatter graph suggests there might be a linear relationship (Figure 6.8, page 94).	Use a correlation table to look up the value of r_s. If the value of r_s is greater than the r_s value for a probability of 0.05, you can state there is a significant correlation between the two values.

Table 6.1 The use of different statistical tests

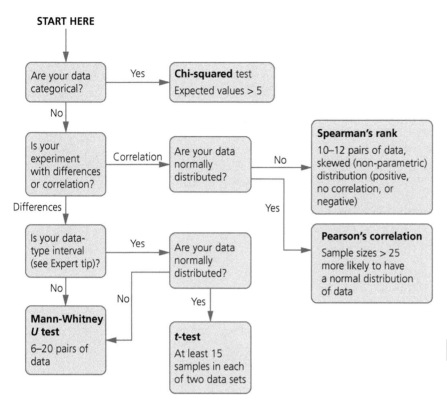

Figure 6.7 Key for identifying the correct statistical test

The application of the *t*-test has been described previously (pages 91–2). You need to be able to carry out the chi-squared test as part of your IB Biology syllabus, and so this will be explained during your course. The other tests might be useful in your IA, although you will not be expected to use or calculate them in an examination. In addition, many calculators are programmed to carry out statistical tests, and computers will run spreadsheet programs with statistical tests programmed in (see Chapter 7, page 105). Dedicated statistical software is available too. The following sites can be used to carry out statistical tests:

- Mann-Whitney U: http://www.socscistatistics.com/tests/mannwhitney/

- *t*-test: http://www.physics.csbsju.edu/stats/t-test_bulk_form.html

- Spearman's rank: http://mathematics.laerd.com/maths/spearmans-rank-order-correlation-calculator.php

- Chi-squared: http://www.socscistatistics.com/tests/chisquare2/Default2.aspx

- Pearson's correlation: http://www.socscistatistics.com/tests/pearson/Default2.aspx

Expert tip

Students are often concerned about the use of statistics to analyse data. The mathematics that underpins statistics can be complex and difficult to understand. However, when analysing results of practical investigations, it is more important to show an appreciation of what the statistics can do or show, rather than how they work.

■ Correlation coefficient

The Pearson correlation coefficient (*r*) measures the strength and the direction of a linear relationship between two variables. The value of *r* lies between −1 and +1 (Figure 6.8). The + and − signs are used to indicate positive linear correlations and negative linear correlations, respectively.

Positive correlation: If *x* and *y* have a strong positive linear correlation, *r* is close to +1. An *r* value of exactly +1 indicates a perfect positive fit. Positive values indicate a relationship between *x* and *y* variables such that as values for *x* increase, values for *y* also increase.

Expert tip

Interval data = data that involve counts; categorical data = data that can be divided into categories.

Expert tip

When researching the techniques to use in your experimental design (methodology) for an IA you should also identify the statistical tests you will use later to analyse the raw data.

Expert tip

Do not choose an investigation solely on the basis that it will involve a statistical test you know and are confident of carrying out. Choose an investigation based on biology you are interested in.

Negative correlation: If *x* and *y* have a strong negative linear correlation, *r* is close to −1. An *r* value of exactly −1 indicates a perfect negative fit. Negative values indicate a relationship between *x* and *y* such that as values for *x* increase, values for *y* decrease.

If there is no linear correlation or a weak linear correlation, *r* is close to 0. A value near zero means that there is a random, non-linear relationship between the two variables.

A perfect correlation of ±1 occurs only when the data points all lie exactly on a straight line. If *r* = +1, the slope of this line is positive; if *r* = −1, the slope of this line is negative.

Examiner guidance

r is a dimensionless quantity; that is, it does not depend on the units used.

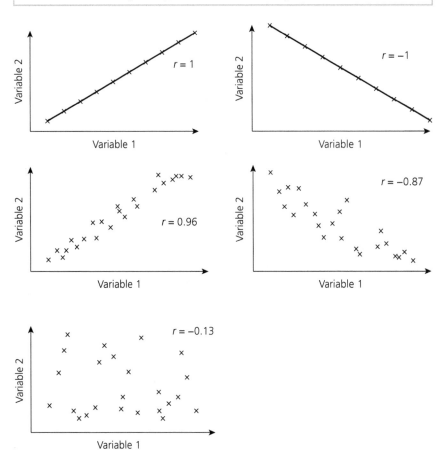

Figure 6.8 The strength of the correlation in the scatter graphs is the correlation coefficient which extends from +1 to −1

A causal relation between two variables exists if changes in one variable cause changes in the other. The first variable can be called the cause and the second the effect. A correlation between two variables does not imply causation.

Further statistical tests

■ Mann–Whitney *U*

This test is used for investigating the difference between two sets of values which are from a small data set which may not be normally distributed.

The test is carried out in the following way:

Values for the two sets of data are entered into a table:

Data-set 1								
Data-set 2								

Both sets of data are then reorganized in a new table and values ranked. Data are ranked together, from lower to higher values, from both sets of values together rather than independently (as with the Spearman's Rank test – see page 96–8):

	Rank																	
	1	2	3	4	5	6	7	8	9	10	11	12	13	14	15	16	17	18
Data set 1																		
Rank (R_1)																		
Data set 2																		
Rank (R_2)																		

The ranks are totalled for each data set (ie added together):

ΣR_1 = sum of ranks for data set 1

ΣR_2 = sum of ranks for data set 2

U values are then calculated using the following formulas:

U for data set 1:

$$U_1 = (n_1 \times n_2) + \frac{n_2(n_2+1)}{2} - \Sigma R_2$$

where:

n_1 = number of samples in data set 1

n_2 = number of samples in data set 2

U for data set 2:

$$U_2 = (n_1 \times n_2) + \frac{n_1(n_1 + 1)}{2} - \Sigma R_1$$

Check the calculation: $U_1 + U_2 = n_1 \times n_2$ (that is, values for U_1 and U_2 when added together should equal the value of n_1 multiplied by n_2).

The *lower* value of U is selected as the test statistic.

The test statistic is compared against the table of critical values for Mann–Whitney U:

$p = 0.05$		Values of n_2									
		1	2	3	4	5	6	7	8	9	10
	1										
	2								0	0	0
	3					0	1	1	2	2	3
	4				0	1	2	3	4	4	5
Values of n_1	5			0	1	2	3	5	6	7	8
	6			1	2	3	5	6	8	10	11
	7			1	3	5	6	8	10	12	14
	8		0	2	4	6	8	10	13	15	17
	9		0	2	4	7	10	12	15	17	20
	10		0	3	5	8	11	14	17	20	23

Unlike other test statistics, the null hypothesis is rejected if the test statistic is *less than* or equal to the critical value.

Worked example

An ecological investigation tested whether human trampling affects the height of a shingle ridge plant species, viper's bugloss (*Echium vulgare*).

Plants were measured in a trampled site and in an enclosure plot (a fenced-off area of the shingle ridge where there was no access for the public) where there was no disturbance. Plant height was measured using a ruler (in cm).

Hypothesis: There is a significant difference in the height of viper's bugloss in the trampled and untrampled areas.

Null hypothesis: There is no significant difference in the height of viper's bugloss in the trampled and untrampled areas.

Data from both sites were collected and ranked:

n	Height of plant in trampled area/cm	Height of plant in untrampled area/cm	Rank 1 R_1	Rank 2 R_2
1	0.0	51.3	3	12
2	0.0	86.2	3	20
3	31.6	0.0	9	3
4	51.0	67.8	11	16
5	52.2	74.2	13	18
6	25.9	71.0	7	17
7	0.0	21.4	3	6
8	31.1	55.7	8	14
9	0.0	66.7	3	15
10	46.3	85.6	10	19
		Sum of ranks	70	140

U values for each of the two sets of data were calculated:

$$U_1 = n_1 \times n_2 + \frac{n_2(n_2 + 1)}{2} - \Sigma R_2$$

$$= 10 \times 10 + \frac{10(10 + 1)}{2} - 140$$

$$= 100 + (55 - 140) = \mathbf{15}$$

$$U_2 = n_1 \times n_2 + \frac{n_1(n_1 + 1)}{2} - \Sigma R_1$$

$$= 10 \times 10 + \frac{10(10 + 1)}{2} - 70$$

$$= 100 + (55 - 70) = \mathbf{85}$$

The lower U value, 15, becomes the test statistic. This value is then compared to the table of critical values at the 5 % significance level. With 10 pairs of data, the critical value is 23.

The null hypothesis is rejected because the test statistic is lower than the critical value (15 < 23). There is therefore a significant difference in the height of viper's bugloss plants in the trampled and undisturbed sites.

■ Spearman's rank

Spearman's rank correlation coefficient is a statistical test which can be used to determine the strength and direction (negative or positive) of a relationship between two variables. The result will always be between 1 (strong positive correlation) and minus 1 (strong negative correlation). In this test, data for two variables are ranked independently, the differences between paired ranked values are calculated, and these values are used to determine Spearman's rank coefficient. If paired values are identically ranked, Spearman's rank will equal 1, indicating a perfect correlation. For values <1, a table of critical values can be used to determine whether a correlation is significant or not.

The equation for Spearman's rank is:

$$r_s = 1 - \frac{6\Sigma d^2}{n(n^2 - 1)}$$

where:

r_s = Spearman's rank correlation coefficient

d = difference between ranks of the two sets of data

d^2 = square of difference between ranks

n = number of pairs of data.

Worked example

An investigation was carried out into the relationship between soil depth and plant height along a coastal shingle ridge succession. Samples were taken every 10 m along a transect perpendicular to the sea. Soil depth and plant height were measured at each location. The tallest plant at each location was selected – the plant was held upright and the height recorded using a tape measure.

Ten pairs of data, of soil depth and plant height, provided sufficient data to carry out the statistical test.

Spearman's rank was used to test whether the correlation between soil depth and plant height is statistically significant.

The following table contains data collected during the investigation. Results show the average values from 5 transects:

	Transect station number									
	1	2	3	4	5	6	7	8	9	10
Average plant height/cm ± 0.1 cm	24.0	31.0	12.5	44.0	33.0	137.0	242.0	200.0	324.0	250.0
Average soil depth/cm ± 0.1 cm	0.0	2.0	14.3	19.9	19.0	35.0	40.0	40.5	81.5	24.0

Table 6.2 Data from a shingle ridge succession investigating the effect of soil depth on plant height

Station 1 is nearest the sea and 10 furthest from it.

Null hypothesis: There is no significant correlation between soil depth and plant height.

Spearman's rank is calculated in the following way:

$$\text{Spearman's rank} = r_s = 1 - \frac{6\Sigma d^2}{n(n^2 - 1)}$$

where d = the difference in ranks and n = number of pairs of data.

Analysis table for Spearman's rank:

n	Soil depth/cm	Plant height/cm	Rank of soil depth	Rank of plant height	Rank difference (d)	(Difference)² (d^2)
1	0.0	24.0	1	2	1	1
2	2.0	31.0	2	3	1	1
3	14.3	12.5	3	1	2	4
4	19.9	44.0	5	5	0	0
5	19.0	33.0	4	4	0	0
6	35.0	137.0	7	6	1	1
7	40.0	242.0	8	8	0	0
8	40.5	200.0	9	7	2	4
9	81.5	324.0	10	10	0	0
10	24.0	250.0	6	9	3	9
$n = 10$					$\Sigma d^2 =$	20
					$6\Sigma d^2 =$	120
					$n(n^2 - 1) =$	990

Table 6.3 Calculating Spearman's rank for plant height and soil depth data

Spearman's rank, $r_s = 1 - \dfrac{120}{990}$

Therefore, $r_s = 0.88$.

When $n = 10$, the critical value for Spearman's rank, at 5 % probability level = 0.648 (Table 6.4).

n	Critical value of r_s
7	0.786
8	0.738
9	0.683
10	0.648
12	0.591
14	0.544
16	0.506
18	0.475
20	0.450
24	0.409
30	0.364

Table 6.4 Table of critical values for Spearman's rank correlation coefficient at $p = 0.05$ level

As the r_s value of 0.88 is higher than the critical value of 0.648, the result is significant at the 5 % probability level. The null hypothesis can be rejected.

There is a significant correlation between plant height and soil depth at $p = 0.05$.

■ Chi-squared test

The chi-squared (χ^2) test is used to examine data that fall into discrete categories. It tests the significance of the deviations between numbers observed (O) in an investigation and the number expected (E). The measure of deviation, known as chi-squared, is converted into a probability value using a chi-squared table. In this way, it can be decided whether the differences observed between sets of data are likely to be real or, alternatively, obtained by chance.

Chi-squared is calculated using the equation:

$$\chi^2 = \Sigma \frac{(O - E)^2}{E}$$

where:

O = observed result

E = expected result.

Worked example

An ecologist may want to investigate whether two species tend to be found together or not. Species that tend to be located together may share similar microhabitat requirements – this gives the ecologist useful information about the organisms involved. A statistical test – the chi-squared test – can be used to assess whether there is an association between two species. If the species are non-motile (ie sedentary) then quadrats can be used to sample the organisms.

Example – Testing the association between two moorland plants

This example examines whether the moorland species bell heather (*Erica cinerea*) and common heather, also known as ling (*Calluna vulgaris*), tend to occur together. Moorlands are upland areas with acidic and low-nutrient soils, where heather plants dominate. Heathers have long woody stems, gow in dense clumps, and have colourful bright flowers. The null hypothesis in this example would be that there is no statistically significant association between bell heather and ling in an area of moorland, that is, their distributions are independent of each other.

In order to sample the two species, the presence or absence of each species was recorded in each of 200 quadrats. The quadrats were located at random on a 100 m by 100 m area of moorland. Two 100 m tapes were placed at right angles to each other. Quadrat locations were chosen by using a random-number generator, which gave two random numbers between 1 and 1 000 for each location – the numbers 548 and 889, for example, meant that a quadrat was located 54.8 m along the bottom tape and 88.9 m along the side tape.

Observed results

	Bell heather present	Bell heather absent	Total
Ling present	89	45	134
Ling absent	31	35	66
	120	80	200

Table 6.5 Observed results – the distribution of ling and bell heather

Expected results

Assuming that the two species are randomly distributed with respect to each other, the probability of ling being present in a quadrat is:

$$\frac{\text{column total}}{\text{total number of quadrats}} = \frac{134}{200}$$
$$= 0.67$$

Similarly,

$$\text{probability of bell heather being present in a quadrat} = \frac{120}{200}$$
$$= 0.60$$

The probability of both species occurring together, assuming random distribution of the two species, is: $0.60 \times 0.67 = 0.40$. The number of quadrats in which both species can be expected is therefore $0.40 \times 200 = 80$.

Having worked out the expected number of quadrats where the species are found together, other expected values can be calculated by subtracting from the totals. For example, the expected number of quadrats with bell heather but no ling is $120 - 80 = 40$. Expected values follow the assumption that totals for each row and column do not change, because the relationship shown by the data is assumed to represent the true relative frequency of each species. Full expected results are:

Expected results

	Bell heather present	Bell heather absent	Total
Ling present	80	54	134
Ling absent	40	26	66
	120	80	200

Table 6.6 The full expected results

The calculated values can be checked by using the ratios represented in the table of observed results. For example, the expected number of quadrats where there is no ling and no bell heather can be calculated as follows:

$$\text{Probability of no ling in a quadrat} = \frac{66}{200} = 0.33$$
$$\text{Probability of no bell heather in a quadrat} = \frac{80}{200} = 0.40$$
$$\text{Probability of neither species in a quadrat} = 0.33 \times 0.40 = 0.13$$

Number of expected quadrats with neither species present = $0.13 \times 200 = 26$ [this figure agrees with the estimated value in the table]

Statistical test

Observed and expected results are therefore:

		Bell heather present	Bell heather absent	Total
Ling present	Observed	89	45	134
	Expected	80	54	
Ling absent	Observed	31	35	66
	Expected	40	26	
		120	80	200

Table 6.7 Observed (O) and expected (E) distribution of ling and bell heather

Chi-squared is calculated using the formula:

$$\chi^2 = \Sigma \frac{(O - E)^2}{E}$$

chi-squared in this example $= \frac{(89 - 80)^2}{80} + \frac{(45 - 54)^2}{54} + \frac{(31 - 40)^2}{40} + \frac{(35 - 26)^2}{26}$

$= 1.01 + 1.50 + 2.03 + 3.12$

$= 7.66$

To find whether this result is statistically significant or not, the value must be compared to a critical value. To locate the critical value, you need to calculate the appropriate degrees of freedom. Degrees of freedom = (number of columns – 1) × (number of rows – 1), and so in this case = (2 – 1) × (2 – 1) = 1.

Degrees of freedom	0.05 level of significance
1	3.84
2	5.99
3	7.81
4	9.49

Table 6.8 Critical values for the chi-squared (χ^2) test at $p = 0.05$ level

The chi-squared value of 7.66 is larger than the critical value of 3.84, for 1 degree of freedom, at the probability level of $p = 0.05$ (the 5 % probability level).

The null hypothesis is therefore rejected, and the hypothesis is accepted, which is that there is a statistically significant association between bell heather and ling in an area of moorland. The distributions of the two species are not independent of each other and the distribution of one species is associated with the distribution of the other. Because the species are found together more frequently than expected, and found on their own less frequently than expected, the data indicate that there is a positive association between ling and bell heather. This suggests that the two species share a common microhabitat, or are influenced by similar biotic or abiotic factors.

Databases

Determining the differences in base sequence of a gene in two species using the Gene Bank database

Use the steps outlined below to compare the nucleotide sequences of the cytochrome c oxidase gene of humans with that of the Sumatran orang-utan. This gene codes for a large transmembrane protein complex (cytochrome c oxidase) which is the last enzyme in the electron transport chain of mitochondria, located in the inner mitochondrial membrane. You will need to use the National Center for Biotechnology Information web services, which provides access to biomedical and genomic information.

Procedure

1   You will need two websites. Get them ready in two tabs in your browser:

■   A: http://www.ncbi.nlm.nih.gov/gene

■   B: http://blast.ncbi.nlm.nih.gov/Blast.cgi

2   Go to website A:

■   Type 'human cytochrome c oxidase' in the search bar and click 'Search'.

■   Click on the blue link for COX6B1 in the search results.

3   This page gives more information about the gene.

■   Scroll down until you find NCBI Reference Sequences (RefSeq). This is close to the bottom of the web page.

■   Click on the first link under the heading 'mRNA and Protein(s)': NM_001863.4. This will give the nucleotide sequence for the gene.

4   Click on the FASTA link at the top of this page. This will give you the sequence of the nucleotides in the mRNA.

■   Copy the whole sequence to your clipboard (ensure you only highlight the nucleotide sequence letters, ATCG, not any of the surrounding text).

5   Go to website B on your second tab:

■   Click on 'Nucleotide BLAST'.

■   Paste your nucleotide sequence into the large box at the top.

■   In the 'Choose Search Set' box below, select 'others' next to Database (this will include comparisons with other species).

■   Scroll down and click on **BLAST**; this will compare the human cytochrome c oxidase sequence with the equivalent sequence in other species.

6   Scroll down to the descriptions and find *Pongo abelii* (Sumatran orang-utan);

■   What percentage match to the human sequence does it have? (Use the 'Ident' column – 96 %).

■   Click on the species link or scroll down to the alignments and compare the gene sequences base by base. What types of mutation can you see (substitutions and deletions)?

7   Scroll back to the top.

■   Click on 'distance tree of results'. An evolutionary tree (**cladogram**) will appear in a new tab. You can change this to 'slanted' (click 'Tools', then 'Layout' and select 'Slanted Cladogram') to get a more familiar diagram. This will show the evolutionary relationships between of various species based on the cytochrome c oxidase gene.

Pongo abelii cytochrome c oxidase subunit Vib polypeptide 1 (ubiquitous) (COX6B1), mRNA
Sequence ID: ref|NM_001131741.1| Length: 536 Number of Matches: 1
▷ See 1 more title(s)

Range 1: 1 to 487 GenBank Graphics   ▽ Next Match ▲ Previous Match

Score	Expect	Identities	Gaps	Strand
797 bits(431)	0.0	469/487(96%)	4/487(0%)	Plus/Plus

```
Query  106  TTCCGCTTCCTGTCCGACTGTGGTGTCTTTGCTGAGGGTCACATTGAGCTGCAGGTTGAA  165
            ||||||||||| || ||||||||||||||||||||||||||||||||||||||||||| |
Sbjct  1    TTCCGCTTCCCGTGCGACTGTGGTGTCTTTGCTGAGGGTCACATTGAGCTGCAGGTTGCA  60

Query  166  TCCGGGGTGCCTTTAGGATTCAGCACCATGGCGGAAGACATGGAGACCAAAATCAAGAAC  225
            ||||||||||||||||||||||||||||||||||||||||||||||||||| ||||||||
Sbjct  61   TCCGGGGTGCCTTTAGGATTCAGCACCATGGCGGAAGACATGGAGACCAAACTCAAGAAC  120

Query  226  TACAAGACCGCCCCTTTTGACAGCCGCTTCCCCAACCAGAACCAGACTAGAAACTGCTGG  285
            |||||||| |||||||||||||||||||||||||||||||||||||||| |||||||||||
Sbjct  121  TACAAGACTGCCCCTTTTGACAGCCGCTTCCCCAACCAGAACCAGACCAGAAACTGCTGG  180

Query  286  CAGAACTACCTGGACTTCCACCGCTGTCAGAAGGCAATGACCGCTAAAGGAGGCGATATC  345
            ||||||||||||||||||||||||||||||||||||||||||||||||||||||||||||
Sbjct  181  CAGAACTACCTGGACTTCCACCGCTGTCAGAAGGCAATGACCGCTAAAGGAGGCGATATC  240

Query  346  TCTGTGTGCGAATGGTACCAGCGTGTGTACCAGTCCCTCTGCCCCACATCCTGGGTCACA  405
            ||||||||||||||||||||||||||||||||||||||||||||||||||||||||||||
Sbjct  241  TCTGTGTGCGAATGGTACCAGCGTGTGTACCAGTCCCTCTGCCCCACATCCTGGGTCACA  300

Query  406  GACTGGGATGAGCAACGGGCTGAAGGCACGTTTCCCGGGAAGATCTGAACTGGCTGCATC  465
            ||||||||||||||||||||||||||||||||||||||||||||||||||||||||| ||
Sbjct  301  GACTGGGATGAGCAACGGGCTGAAGGCACGTTTCCCGGGAAGATCTGAACTGGCTGCGTC  360

Query  466  TCCCTTTCCTCTGTCCTCCATCCTTCTCCCAGGATGGTGAAGGGGGACCTGGTACCCAGT  525
            ||||||||| ||||||||| |||||||||||||||||||||||||||| | ||||||| ||
Sbjct  361  TCCCTTTCCTTTGTCCTCCGTCCTTCTCCCAGGATGGTGAAGGGGGATGTAGTACCCCGT  420

Query  526  GATCCCCACCCCAGGATCCTA----AATCATGACTTACCTGCTAATAAAAACTCATTGGA  581
            |||||||||||| |||||||||        |||||||||||||||||||||||||||||||||||
Sbjct  421  GATCCCCACCCCGGGATCCTAAATCAATCATGACTTACCTGCTAATAAAAACTCATTGGA  480

Query  582  AAAGTGA     588
            |||||||
Sbjct  481  AAAGTGA     487
```

Figure 7.1 Screenshot comparing two nucleotide sequences for the same gene (cytochrome c oxidase) from two species (human and Sumatran orang-utan)

▥ Use of databases to identify the locus of a human gene and its polypeptide product

Background information

Genes are located at a specific position on a chromosome (its locus). Online databases can be used to locate the locus of specific genes in the human genome. The database also provides information about the function of the gene (for example, its polypeptide product).

Procedure

1 Access the *Online Mendelian Inheritance in Man* website: http://omim.org/.

2 Go to the Gene Map search engine: http://omim.org/search/advanced/ geneMap.

3 Enter the name of a gene into the search engine (access a list of genes here: http://en.wikipedia.org/wiki/List_of_human_genes). A box will appear with information about the gene, including the chromosome it is found on ('Location', the number before the colon), its locus and details of its function.

4 Alternatively, you can enter the number or name of a human chromosome (autosomal (1–22) or the sex chromosomes (X or Y)). A complete sequence of gene loci for the chromosome will be displayed.

Spreadsheets: Analysing data with Excel

Computer spreadsheets, such as Excel, provide the means to analyse quickly large amounts of data, and to present outcomes as graphs. This section outlines how Excel can be used to analyse data from a biological investigation.

Performing a calculation

Use an equals sign (=) in front of the formula. For example, if you want to sum the numbers in cells A1 through A12, then in an empty cell type =SUM(A1:A12). If you want to write a formula for the expression $\dfrac{x^2 y}{z + w}$ where x is in cell A1, y is in cell A2, z is in cell B3, and w is in cell B5, you could write the formula in cell C2 as =A1^2 * A2/(B3 + B5).

Copying a formula

If you have typed a formula in cell B1 and you want it in cells B2 through B10, position the cursor in the lower right-hand corner of selected cell B1 until the cursor becomes a black cross. Click and drag from cells B2 through B10. Note that any cells written in the original formula (in cell B1) will be shifted when the formula is copied, unless the cell references are written using $ symbols, such as A2. For instance, if B1 contains =2*A1 and you copy B1 to B2, the formula in B2 will be =2*A2.

Built-in functions in Excel

Excel has a number of built-in functions that can be used to process data and carry out statistical analysis (see page 105). The function wizard, *fx*, on the toolbar will bring up a dialogue box where all the built-in functions are listed.

Excel formula	Description of the formula
=SUM(A2:A5)	Find the sum of values in the range of cells A2 to A5.
=COUNT(A2:A5)	Count the number of numbers in the range of cells A2 to A5.
=COUNTIF(A1:A10,100)	Count cells equal to 100.
=COUNTIF(A1:A10,'>30')	Count cells greater than 30.
=ABS(A2)	Find the absolute value of the number in cell A2.
=SQRT(A2)	Find the square root of the number in cell A2.
=EXP(A2)	Find the exponential of the number in cell A2, ie *e* raised to the power of the value in cell A2.
=LN(A1)	Find the natural logarithm of the number in cell A1.
=LOG(A1)	Find the logarithm (to the base 10) of the number in cell A1.

Table 7.1 Built-in functions in Excel

Graphing in Excel

Select the block of cells in Excel containing the data to be plotted (which might include headings). The *x*-axis data column should always be to the left of the *y*-axis data column. For example, place *x*-axis data in column A and *y*-axis data in column B.

Click on the Insert tab and choose the graph type from the Charts area, usually scatter if the data are continuous. Graphs created on a separate Excel sheet can easily be copied or pasted into a Word document.

Graphs embedded into an Excel worksheet can be edited even after they have been inserted. Experimental data will show scatter due to random errors in the measurement. You can add a trendline by right clicking on a data point and selecting Add Trendline. If the data lie on an approximately straight line and a linear relationship is expected then select linear in Trendline Options. Check the relevant boxes to display the equation of the line (regression equation) and the correlation value.

Simulations with Excel

Computer modelling is the process of creating computer-based representations of the structure and interactions of biological systems. It is used to understand the underlying causal factors of a biological phenomenon, to allow prediction and to bring theory and experiment together. The main issues in developing a computer model are deciding on its assumptions and simplifications, approximating but simplifying real-world conditions, and introducing its limitations.

Models can be deterministic or stochastic. Deterministic models are calculated with fixed probabilities. Stochastic models use a random number generator to create a model with variable outcomes. Using the Random function in a spreadsheet (to simulate the rolling of dice), stochastic models of biological phenomena can be created and explored, and data from a number of simulations can be collected and analysed.

Expert tip

Excel provides two functions for generating random numbers (remember that a function requires '=' in front of it):

RAND() – generates a random number between 0 and 1

RANDBETWEEN(*a, b*) – generates a random integer between *a* and *b*

These functions are volatile, meaning that every time there is a change to the worksheet their value is recalculated and a different random number is generated.

RANDBETWEEN only generates integer values. If you want a random number that could be any decimal number between *a* and *b*, use the following formula instead:

=*a*+(*b*–*a*)*RAND()

Hardy–Weinberg simulation using Excel

The Hardy–Weinberg equation shows the relationship between gene frequencies and genotype frequencies in random mating populations.

$(p + q)^2 = p^2 + 2pq + q^2$

p = frequency of dominant allele A

q = frequency of recessive allele B

p^2 = frequency of individuals with genotype AA

q^2 = frequency of individuals with genotype BB

$2pq$ = frequency of individuals with genotype AB

The Hardy–Weinberg principle states that the amount of genetic variation in a population (allele and genotype frequencies) will remain constant from one generation to the next in the absence of evolutionary forces such as mutation and natural selection.

In cell D2, enter a value for the frequency of the A allele. This value should be between 0 and 1.0. Because all of the alleles in the population are either A or B for a given trait, the Hardy–Weinberg equation $p + q = 1$ applies. Enter =1–D2 in cell D3.

In cells A2 and A3 enter the labels: *p*=freq allele A and *q*=freq allele B. In cell B8, enter the function =RAND() (which generates random numbers between 1 and 0 but is removed later). If you are using a PC, try pressing the F9 key several times to force recalculation. On a Mac highlight the formula and press return.

In cell E5 enter the formula =IF(RAND()<=D\$2,"A","B") to generate a random number, compare it with the value of *p*, and then place either an *A* gamete or a *B* gamete in the cell. Enter the same formula in cell F5. Then copy these two formulas in cells E5 and F5 down for 25 rows to represent gametes that will form 30 offspring for the next generation.

The zygote is a combination of the two randomly selected gametes (A and B). The two gametes (represented as strings) need to be concatenated: in cell G5 enter =CONCATENATE(E5,F5) and then copy the formula down the spreadsheet.

The next columns in the sheet, H, I and J, are used to maintain a tally of the numbers of each zygote's genotype. Enter =IF(G5="AA",1,0) into cell H5, so if the value in cell G5 is *AA*, the formula will put a 1 in the cell; if not, the formula will put a 0. Enter a similar formula =IF(G5="BB",1,0) into cell J5.

In cell I5 enter the nested function =IF(G5="AB",1,(IF(G5="BA",1,0))). If the value in cell G5 is exactly BA, the formula will put a 1 in cell I5; if it is neither, then the formula will put a 0 in this cell. Copy these three formulas down for the rows in which gametes were produced.

Calculate the sum of the AA genotypes in column H by entering the following formula in cell H35 =SUM(H5:H30). Calculate the sum of the AB and BB genotypes in columns I and J using the same formula in cells I25 and J25, respectively (changing the 'H's to 'I's and 'J's as required).

Descriptive statistics with Excel

Spreadsheet programs, such as Excel, have a number of built-in (and add-in) statistical functions. The term average refers to the 'centre of distribution', which refers to the mean, the mode or the median, which measure the central tendency of a sample.

Descriptive statistics	Excel formula	Comment
Arithmetical mean or mean	=AVERAGE(range)	You should never calculate a mean from numbers that are means.
Median	=MEDIAN(range)	If there is an even number of values, this is the mean of the middle two values.
Mode	=MODE(range)	The value that occurs most frequently.

Table 7.2 Statistical functions available in Excel

The use of these Excel formulas is illustrated in Table 7.1. In many cases the quantities measured are continuous variables and will show a normal (Gaussian) distribution, and so the arithmetical mean is the most appropriate descriptive statistic to use.

If the mean of a sample is calculated then a measure of the variation, dispersion or spread of the data should also be calculated.

The range is given by the Excel formula: =MAX(range)−MIN(range), where MAX returns the largest value in a range and MIN the smallest. This is the simplest but least useful function since it is a relatively crude measure that does not take the other (intermediate) values into account.

The variance (of a population sample) is given by the Excel formula: =VAR(range), but this is not useful as a descriptive statistic since it is not in the same units as the measurements.

The standard deviation (of a population sample) is given by the Excel formula: =STDEV(range). The standard deviation (SD; can also be represented as Greek letter sigma, σ) is the square root of the variance and gives a good indication of the variability or 'spread' of a set of data around the mean. 68.2 % of the results are within 1SD of the mean, 95.4 % are within 2SD of the mean and 99.7 % are within 3SD of the mean (see Chapter 6, pages 85–7).

Expert tip

Standard deviation is not the best statistic to use when comparing different sets of data, especially if the data sets are of different sizes.

Examiner guidance

If the numbers you are analysing represent an entire population, rather than a sample, then use the VARP and STDEVP functions to calculate variance and standard deviation.

The standard error of the mean is given by the formula: =STDEV(range)/SQRT (COUNT(range)). This gives an indication of the confidence of the mean. The 95% confidence interval is given by the formula: =CONFIDENCE(0.05, STDEV(range), COUNT(range)).

The value of 0.05 is used to give the 95% (0.95) confidence interval, which means there is a 95% probability that the real or true mean lies within ± confidence interval from the measured mean (see Figure 7.2; see also Figure 6.3). It indicates the percentage number of times that the calculated interval actually contains the true population mean when the process of taking n samples is repeated a large number of times. However, this means that there is still a 1 in 20 probability that the calculated confidence interval will not contain the true population mean. The upper and lower limits of this range are called the confidence limits and can be shown as error bars on line graphs or bar charts.

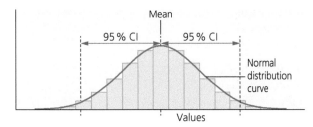

Figure 7.2 The normal distribution curve and 95% confidence intervals

Data logging

Data loggers are electronic devices that record data over time, using sensors (Figure 7.3). There are many different sensors, each used to measure different abiotic variables (see Chapter 5, pages 71–2), such as carbon dioxide concentration, dissolved oxygen concentration, pH and temperature. Other data loggers measure, for example, breathing rate.

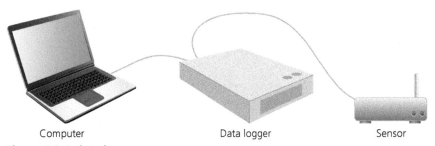

Computer   Data logger   Sensor

Figure 7.3 A data logger

Different makes of data logger are available, but those produced by Vernier are widely used. Vernier lab manuals have basic modules which can be extended for the purposes of IAs (https://www.vernier.com/products/packages/advanced-biology/labq2/).

Data loggers allow high-resolution data to be produced, with high levels of precision. They usually have a greater degree of sensitivity than other digital equipment, recording data more accurately to a greater number of decimal places. They allow data to be collected continually over the short term (that is, minutes or hours) or long term (weeks or months).

If your school has data loggers, spend some time exploring how they work, using the manuals available.

Many apps for smartphones allow you to record data (see below), if you do not have access to data loggers at your school or college.

Use of smartphones

Your smartphone is a powerful computer, and contains complete operating systems which allow software programs to operate on them ('apps'). There are numerous types of app available for smartphones, many with science-specific applications. Up to 60 % of them are free and many others are available at reasonable prices.

How to obtain apps

Your smartphone will already have apps installed on it. Other apps can be downloaded from online stores. Online search engines, such as Google, enable you to locate suitable apps, for example, by searching for 'biology apps for school/college'. Searching for 'Google play' brings up Android apps.

Some organizations have developed their own apps, for example:

- American Association for the Advancement of Science (AAAS): http://sciencenetlinks.com/collections/science-apps

- National Aeronautics and Space Administration (NASA): www.nasa.gov/connect/apps.html#.U34C6fldV8E

Recommended apps can be found by reading online reviews, for example:

- https://www.sellcell.com/blog/five-data-logging-apps-for-schools-and-colleges/

- https://www.tomsguide.com/us/pictures-story/962-best-science-apps.html

- www.wired.com/wiredscience/2008/07/20-iphone-apps

Apps for biology

New apps are being developed all the time. Here are some existing apps that you might find useful.

- brunalab.org/apps: A wide-ranging list of apps for biology

- https://identify.plantnet-project.org/: Pl@ntNet is a tool to help you identify plants using pictures

- https://play.google.com/store/apps/details?id=com.cosalux.welovestars: For estimating the brightness of the night sky as part of a project on light pollution

- www.usanpn.org/nn/mobile-apps: For data entry in the field

- www.noisetube.net/: Monitors any noise pollution anywhere, but app can be used in labs as it measures sound in decibels

- https://itunes.apple.com/us/app/expedition-white-shark/id488682903: 'Expedition white shark' allows you to follow tagged great white sharks

- www.inaturalist.org: Allows you to record the location of a species in the field

Using smartphones to record data

Sensors are built into smartphones for specific purposes, but particular apps can use the sensors as external measuring instruments for investigations. Other apps record geographical data, for example, iNaturalist provides a means of recording the geolocation of a species in ecological field studies.

Apps are available on smartphones that allow biological principles to be investigated, such as **evolution** by natural selection: https://itunes.apple.com/us/app/evolutionary-biology/id513464425?mt=8

Your smartphone provides many opportunities to record biological data. Here are some ideas for you to try.

Expert tip

If you use your smartphone to record data for an IA, you need to ensure that the measurements taken are accurate enough to be used in quantitative experiments. Ask your teacher if you are unsure.

Common mistake

Some apps display graphs that do not have proper axes showing independent and dependent variables, and that exclude important features such as units.

Observing cells: converting smartphones into 'CellCams'

Photographs have traditionally been taken of microscope images (photomicrographs) using expensive camera-mounted apparatus. Smartphones make it possible to take high-quality images without the need for such specialized apparatus. Light passing through the lens can be captured using a properly aligned smartphone camera. It is difficult to take photographs using a hand-held smartphone, because any slight variation in the angle of the phone to the light-beam will render the image invisible or partially obscured. However, simple methods can be used to attach the smartphone to the eyepiece of the microscope, so that images can be taken. Such adaptors enable you to create your own 'CellCam'.

Modify a toilet-paper cardboard tube (4.5 cm diameter and 4 cm in length) by adding adhesive foam weather-strip (1–2 cm wide) to the inside, so that the tube fits securely around the eyepiece of the microscope. When using the CellCam, you should remove any protective casings. For a mid-mounted smartphone camera, attach the tube to the camera using ice-cream sticks secured either side of the tube using elastic bands (Figure 7.4a). For a corner-mounted camera, cut a slit in one edge of the tube and place it around the camera (Figure 7.4b).

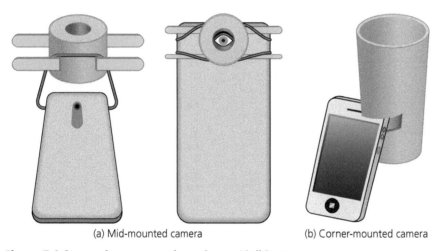

(a) Mid-mounted camera   (b) Corner-mounted camera

Figure 7.4 Converting a smartphone into a 'CellCam'

The CellCam can be used to capture images or video of specimens (Chapter 3). Bring the image into focus in the usual way, then attach the camera to capture the image. Place the camera in the centre of the adaptor and slowly push the tube onto the eyepiece until the field-of-view fills the screen. By keeping one hand on the adaptor and the other supporting the phone, you can keep the the image centred. The camera will automatically adjust the exposure so that a clear image is produced.

The size of specimens can be estimated using the field-of-view. If a plastic ruler is first placed on the microscope stage, so that the millimetre marks are clearly shown, a photo can be taken and the image used to measure the space between single millimetre increments. These measurements can be used to determine the size of specimens viewed under the microscope. For example, if a 1.0 mm increment measures 10 mm on the smartphone screen, then the magnification is ×10. If a human hair is photographed and has a screen diameter of 0.8 mm, the actual diameter of the hair will be 0.8/10 = 0.08 mm. The magnification and associated calibration will change from one objective lens to another, so the procedure must be repeated for each objective lens (that is, low, medium and high power).

▦ Investigating plant movement

Smartphones can be set up to record time-lapse photography. Apps are available to show movements that are normally too slow to see, such as Lapse It: http://www.lapseit.com/.

This software can be used to record plant movements, such as phototropic and geotropic responses to light and gravity, respectively. Plants such as *Arabidopsis* demonstrate, for example, geotropic responses within a two-hour lab session. *Brassica* plants can also be used. Plants can be grown by scattering a few seeds over wet soil in small plastic pots arranged in rectangular rows, covered by a clear plastic sheet to keep humidity high during germination. One week after germination, the plastic sheet can be removed and the plants bottom-watered by submerging the pots so that water enters the soil by capillary action. Fertilizer can be added to the water to enhance growth.

After a few weeks, the plants should be large enough to study. Place individual plants in their own pots so they can be studied independently. Flowering stems from 2 to 10 cm in length are suitable for investigations. Geotropic responses can be observed by turning pots 90° from the main vertical axis, and phototropic responses can be observed by exposing the plants to unidirectional light.

To make a time-lapse recording:

- Using Lapse It (see above), set the frame interval to 2 and the time scale to 2 minutes.

- Ensure that the stem is fully visible to the smartphone camera. The phone can be supported using available apparatus, such as test tube racks.

- Place a piece of paper or card behind the plant (any colour but green; blue, purple and white provide a good contrast to the plant).

- To initiate a new recording, choose 'New Capture'.

- To start recording, touch the red 'Capture' button.

- Avoid disturbing the plant during the recording session.

- 'Stop' the programme, and play the video to view the results.

- Recordings are placed into 'Gallery' and need to be 'Rendered'.

- Rendered videos are then copied to 'Photo Roll' and can be downloaded.

Ideas for investigations

Different mutants of *Arabidopsis thaliana* are available. The differential response of these varieties can be studied with respect to either phototropic or geotropic responses, using time-lapse photography.

Simulations

A computer simulation can be used to carry out IA investigations, provided it is interactive and open-ended. In order to use a simulation for an IA, you must be able to control variables and design a method that allows the control of each identified variable that affects the dependent variable.

There are many simulations available online. Some of them have PDFs that explain the simulation.

- http://virtualbiologylab.org/: simulations mainly for ecology; two simulations on evolution (including melanistic moths) and one on cell membranes. If you are studying Option C for ecology, the Simpson's reciprocal index of diversity can be used to analyse data from the simulation on stream biodiversity from this website.

- Other useful simulations can be found here: https://phet.colorado.edu/en/simulations/category/biology

- Computer simulations on enzyme action including metabolic inhibition are available, for example: http://www.virtualimage.co.uk/html/enzyme_lab.html

- A simulation of an investigation into the effect of light intensity on the rate of photosynthesis (Chapter 4, pages 63–6) can be found here: http://www.reading.ac.uk/virtualexperiments/ves/preloader-photosynthesis-full.html

---

Ideas for investigations

The peppered moth (*Biston betularia*) provides an example of natural selection in action, and is included as an application in Topic 5 of the IB Biology syllabus. Although it is not possible to study the selection of the peppered and melanic varieties in the wild, there are simulations available online that allow you to do this, for example, http://peppermoths.weebly.com/

From the homepage select 'A bird's eye view of natural selection', which takes you to the simulator. You can select either a polluted forest (with trees covered by soot and no lichen) or unpolluted forest (with trees covered in lichen).

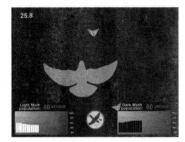

Figure 7.5 Screenshot from peppered moth simulation. The investigator operates a bird predator – moths visible to the operator are 'eaten' by the bird, for a fixed period of time. Relative proportions of light (peppered) moth and dark (melanic) moth are shown. This screenshot shows a polluted forest – a second background, showing an unpolluted forest, is also available.

If you are using the simulation for an IA, you need to manipulate it in some way and adapt it to a specific research question. For example, you could adjust the brightness of your computer screen to investigate how this affects the levels of predation, and selection pressure, on peppered moth populations. You need to ensure the control of variables that might affect your dependent variables – for example, ensure background lighting in the room is constant, and allow the same amount of time for each experiment. You should replicate the investigation and calculate mean results. You could carry out a control experiment by running the simulation under full light intensity. Include screenshots from the simulation in the method and analysis, to clearly explain your methodology and results.

Personal engagement
- Selecting an appropriate investigation
- Personal input and initiative

Writing the
IA report

Communication
- Structure and clarity
- Relevance and conciseness
- Terminology and conventions
- Referencing
- Report format
- Academic honesty

Assessment report

Exploration
- Planning
- Variables
- Background information
- Methodology
- Safety, ethical and environmental issues
- Risk assessments

Analysis
- Recording a presentation of raw data
- Presenting and interpreting biological drawings
- Data processing
- Presenting data – graphing
- Impact of measurement uncertainty
- Interpreting processed data

Evaluation
- Conclusion
- Strengths and weaknesses of the investigation
- Limitations of the data and sources of error
- Improvements and extensions

8 Personal engagement

This criterion assesses the extent to which you engage with the exploration and make it your own. Personal engagement can be demonstrated by addressing personal interests or showing evidence of independent thinking, creativity or initiative in the design, implementation or presentation of the investigation.

Mark	Descriptor
0	The student's report does not reach a standard described by the descriptors below.
1	**The evidence of personal engagement with the exploration is limited with little independent thinking, initiative or insight.** The justification given for choosing the research question and/or the topic under investigation does not demonstrate **personal significance, interest or curiosity**. There is little evidence of **personal input and initiative** in the designing, implementation or presentation of the investigation.
2	**The evidence of personal engagement with the exploration is clear with significant independent thinking, initiative or insight.** The justification given for choosing the research question and/or the topic under investigation demonstrates **personal significance, interest or curiosity**. There is evidence of **personal input and initiative** in the designing, implementation or presentation of the investigation.

© IBO 2014

Table 8.1 Mark descriptors for the Personal engagement criterion

For your IA, you must design your own individual procedure. The subject for an IA can be inspired by an observation, issue or subject area of personal significance. You can demonstrate your personal input and engagement:

■ in your research for background information

■ by persevering while collecting relevant raw data under difficult circumstances – for example, while collecting ecological data in poor weather conditions such as torrential rain

■ in your choice of methods of analysis (Exploration)

■ when establishing the scientific context of the conclusion (Analysis and Evaluation).

When considering this criterion you should bear in mind the following points:

■ Your report should have a statement of purpose.

■ You need to demonstrate the relationship of your research question to/with the real world.

■ The design of the methodology (choice of materials and methods) should show originality.

This criterion is marked using a holistic approach, using the contents of the full report for the Individual Investigation. It will therefore overlap with components of other criteria, for example:

■ Exploration: in the care shown in selecting mathematical and statistical techniques to process the data.

■ Analysis: by your comments concerning the quality of your raw data; and the type of material you refer to in your background research and when discussing your results.

■ Evaluation: through the depth of understanding you show in assessing the limitations of your investigation; through your reflective comments on the improvement and extension of your investigation.

Common mistake

Avoid contriving the personal significance of your investigation, for example, by simply stating 'I have always been interested in ...'.

Expert tip

You cannot use a classic investigation (the prescribed mandatory IB Biology practicals (Table 2 in the Introduction, page xii) or well-known methods in text books) if there is little or no attempt to modify the method. The subject of your investigation must be original in some way. Personal input can be reflected at the simplest level by having completed the investigation, but you cannot expect to score highly if you simply follow a classic experiment. There must be some indication in your report that you have showed commitment to the investigation.

The following guiding questions might help you develop a plan for your Individual Investigation:

- What exactly are you going to investigate?

- Why is it worthwhile or justified to investigate this?

- Why are you personally interested?

- Are there opportunities to show personal engagement?

- What is already known in the biological literature?

- What new biological knowledge is currently being investigated in this area?

- Is your research question answerable within the time available and using the resources in your biology department?

- What method will you use: an 'established' methodology applied to a new topic or a 'new' methodology applied to an established topic?

- Can the investigation be organized into a sequence of experiments?

- How many samples will you be able to analyse? Which planned statistical test and sampling techniques will you use (if appropriate)?

- How many experiments will you conduct? How long will each experiment take?

- Do you have enough time to complete all the experiments in the lessons allocated and in an overall time of 10 hours?

- How will you record and organize your raw data and observations?

- Will you use a data logger and probes to record some raw data?

- What apparatus, instruments or techniques will you use? Do you need chemicals (for example, enzymes)? Are they readily available? Are they safe to use and stable? How should the chemicals be handled and disposed of after practical work? Do you need access to specific software such as BLAST?

- Do you know how to operate the apparatus, such as a respirometer, colorimeter, gel electrophoresis or potometer?

- How will you control and monitor the controlled variables? Can you find secondary data?

- Can some aspects of the investigation be simulated or modelled by computer software?

Look at Figure 3 on page xi. Are you drawn towards any particular concept or topic? Consider Figure 8.1 below, which shows how the biology practicals, introduced in the Introduction and described in Chapters 2 to 5, link to the six key biological concepts and four central themes (the 'levels of integration' in Figure 3). Think about which key concepts, or themes, interest you, and see which practicals are associated with these. Note that some practicals overlap different themes (that is, several can be categorized as both biochemistry and physiology practicals, depending on the emphasis of the investigation).

Justifying your research question

You need to explain why you chose this specific research question and topic. Why is it interesting to you? Be inspired by observations you have made independently or during your IB Biology course. Do not write 'I have always been interested in …'! Your interest needs to be integral to the report and not an 'add on'.

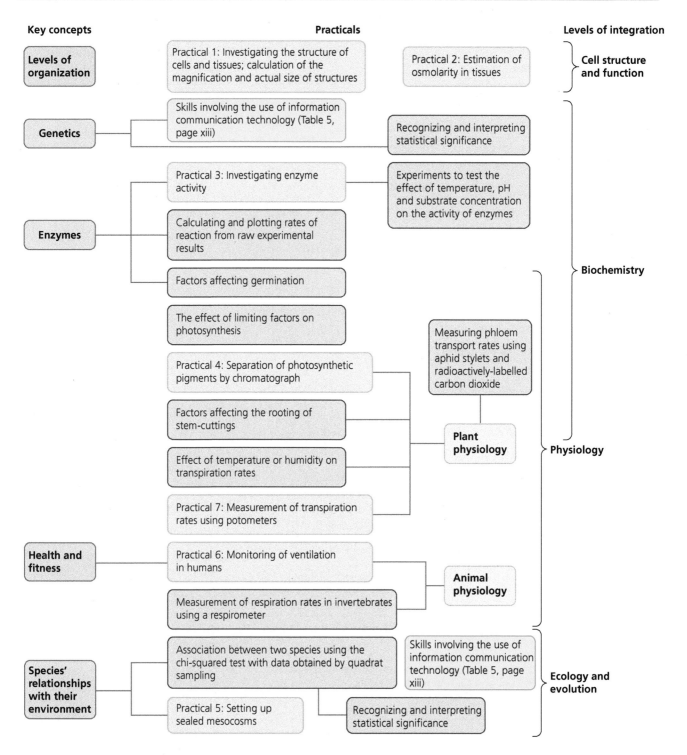

Figure 8.1 Concepts and levels of integration within the IB Biology practicals. Green boxes indicate mandatory practicals; purple boxes indicate practicals listed as 'Skills' in the *IB Biology Guide*; orange boxes indicate ICT skills (see Table 5 in the Introduction, page xiii)

■ Evidence of personal input

There needs to be evidence of independent thinking, personal input and initiative in the design of your investigation and/or in its implementation. This can be evidenced through the level of commitment you show throughout the whole process, including persistence in data collection, design of apparatus, or modification of techniques.

Personal engagement criterion checklist

Creativity, input and initiative

Descriptor	Complete
You demonstrate creativity during the Individual Investigation.	
You have used a suitably modified method to address your specific research question.	
You demonstrate independent thinking during the entire Individual Investigation.	
You demonstrate initiative during the entire Individual Investigation.	

Justification for research question

Descriptor	Complete
You justify why you chose to investigate your research question and include your personal significance, interest and curiosity.	
You discuss the wider importance/impact of the research question.	

9 Exploration

This criterion assesses the extent to which you establish the scientific context for the work, and state a clear and focused research question. You need to show that you have used concepts and techniques appropriate to IB Diploma level. Where appropriate, this criterion also assesses awareness of safety, environmental and ethical considerations.

Mark	Descriptor
0	The student's report does not reach a standard described by the descriptors below.
1–2	The topic of the investigation is identified and a research question of some relevance is **stated but it is not focused.** The background information provided for the investigation is **superficial** or of limited relevance and does not aid the understanding of the context of the investigation. The methodology of the investigation is only appropriate to address the research question to a very limited extent since it takes into consideration few of the significant factors that could influence the relevance, reliability and sufficiency of the collected data. The report shows evidence of limited awareness of the significant **safety,** ethical or environmental issues that are **relevant to the methodology of the investigation.***
3–4	The topic of the investigation is identified and a relevant but not fully focused research question is described. The background information provided for the investigation is mainly appropriate and relevant and aids the understanding of the context of the investigation. The methodology of the investigation is mainly appropriate to address the research question but has limitations since it takes into consideration only some of the significant factors that could influence the relevance, reliability and sufficiency of the collected data. The report shows evidence of some awareness of the significant **safety,** ethical or environmental issues that are **relevant to the methodology of the investigation.***
5–6	The topic of the investigation is identified and a relevant and fully focused research question is clearly described. The background information provided for the investigation is entirely appropriate and relevant and enhances the understanding of the context of the investigation. The methodology of the investigation is highly appropriate to address the research question because it takes into consideration all, or nearly all, of the significant factors that could influence the relevance, reliability and sufficiency of the collected data. The report shows evidence of full awareness of the significant **safety,** ethical or environmental issues that are **relevant to the methodology of the investigation.***

© IBO 2014

* This indicator should only be applied when appropriate to the investigation.

Table 9.1 Mark descriptors for the Exploration criterion.

As well as describing and explaining a research question, this criterion should include background information that provides context and reasons for the investigation. Background information needs to be focused and contain relevant information: superficial or irrelevant material should be avoided. The background will provide a brief overview of the theory and current knowledge, with a special emphasis on the biological literature specific to the Individual Investigation topic. It serves as well to support the argument behind your report, using evidence from that research area.

The independent variable and its range need to be stated and justified, and the dependent variable explained (as well as the processed dependent variable, if appropriate). Preliminary trials may be used to determine the most appropriate values for an independent variable. Introduction of the dependent variable should lead to description of how measurements will be taken. The discussion of controlled variables is needed to demonstrate that you understand and appreciate that other factors might affect values of the dependent variable. Consider whether you can use a control experiment in your investigation (page ix). Scientific names need to be used here and throughout the report.

The variables could be classified and tabulated with an emphasis on explicitly identifying the controlled variables that affect the results. Table 9.2 shows one suggested format for the presentation of information about variables. The example is an investigation determining the effect of temperature on hydrolysis of starch by amylase.

Type of variable	Variable	Method for control	Reason for control
Independent	Temperature	Samples of the enzyme and the substrate will be placed in a thermostatically controlled water bath at a fixed temperature.	N/A
Dependent	Time taken to hydrolyse the starch	The reaction will be monitored by taking samples of the enzyme–substrate mixture at 30-second intervals.	N/A
Controlled	Concentration of starch at start of experiment	A 1 % starch solution will be used for each experiment.	Concentration of starch will affect the rate of reaction, increasing the number of enzyme–substrate complexes produced.
Controlled	Concentration of amylase at start of experiment	A 0.1 % amylase solution will be used for each experiment.	Concentration of amylase will affect the rate of reaction, increasing the number of enzyme–substrate complexes produced.

Table 9.2 A possible format outlining the classification of variables

The method should not be a long list of detailed instructions, but it should explain why certain actions are performed and explain the roles of certain steps. It can be written in continuous prose (as in a scientific paper) or as a list (in bullet points). Any steps or procedures designed to minimize the systematic and random errors in all of your measurements should be clearly described. The method should include the equipment used and a description of how the investigation was carried out. It needs to be written clearly – this will be reflected by marks in both the Exploration and Communication criteria.

You should provide enough detail so that another student could repeat your work without your presence.

This means including information about:

- actual quantities of substances or materials
- how you will measure these quantities
- concentrations of substances
- sizes and precision of apparatus and instruments.

You are allowed to modify a method from a published or online source but you must reference this source.

The report needs to describe and explain the safety, ethics and environmental impact of the investigation (see pages 122–3). As well as identifying potential areas where safety is an issue in your investigation, you also need to explain how each issue will be avoided. If your investigation does not have ethical or environmental impacts then you need to state this and explain why (for example, when conducting a simulation there are no safety or ethical aspects to comment on).

Examiner guidance

Your method should be more than just a list of instructions. It should tell your teacher (and the person moderating your coursework), why you chose certain techniques or apparatus/instruments or why you chose a particular statistical approach to analyse the data. Include any special precautions you took to increase the accuracy, precision or reliability of your raw data.

Expert tip

The method can be written in prose or in the style of a recipe (bulleted/ numbered as a list). Both are acceptable.

Common mistakes

- Weaker reports tend to investigate a topic in which causal relationships are difficult to confirm and a large number of controls are missing, for example, human physiology studies with limited data sets and poorly controlled variables.
- Many candidates lack a sufficiently focused research question. For example, a research question asking how 'different amounts of dissolved sugar will affect cell respiration in yeast' is insufficient because it should state which sugar was used (for example, sucrose) and use a more specific word than 'amount', such as mass, volume or moles. The question should also indicate the range of sucrose concentrations used.
- Some microbiological methods are inappropriate for a school environment and should not be used for an IA. For example, culturing microbes at 37 °C, testing the bactericidal properties of saliva, and culturing bacteria from tooth plaque to test the effectiveness of toothpastes are all unacceptable. Incubation of microbes should be carried out below 30 °C.

Expert tips

- If carrying out a microbiology investigation, make sure you always label cultured plates so that they can be clearly identified. Never open plates for inspection, because this risks contamination or the escape of harmful microbes. Seal Petri dishes with tape, but do not put tape all the way around the dish because this creates anaerobic conditions. Do not assume that the microbes growing on your plates are the strain that was inoculated, even if non-pathogenic strains have been used. At the end of the investigation, sterilize used cultures and dispose of the cultures according to local health and safety regulations.

- The IA should be the product of an individual candidate, and as such excludes most group work because it is impossible, practically and realistically, for all students in a group to completely cover the Personal engagement and Exploration criteria independently, as is required, and still arrive at a shared methodology that would allow them to collect data that can be pooled. In an ecological investigation, students may refer to measurements taken by other students to share some data (for example, measurements of monitored variables), although without a similar sampling design it is difficult to completely pool data.

- If carrying out a human physiology investigation, it is essential that consent forms are obtained from each volunteer.

Examiner guidance

In the same class, it is possible for separate students to work on similar experiments, for example, on rates of reaction of enzymes, as long as they work independently, and on significantly different enzymes, different methods of following the rate of reaction, or different independent variables.

Constructing a hypothesis

A research question *can* be used to formulate a hypothesis and this *might* be a useful inclusion in your background information, but it is not an explicit requirement of the group 4 assessment criteria.

A hypothesis is a testable prediction (ideally quantitative) of how you think an independent variable and a dependent variable are causally related. A hypothesis includes a scientific explanation (often involving a reference to a scientific model) of how the two variables are related.

For example, you might be planning to investigate the relationship between temperature and the rate of fermentation of brewer's yeast. Your hypothesis might be that, as temperature increases, the rate of fermentation is expected to increase exponentially up to an optimum temperature and then decrease rapidly (in a curved manner) to zero.

Examiner guidance

The relationship between temperature and rate of fermentation can be explained by reference to simple collision theory, based on average kinetic energies of particles, and the concept of activation energy, as a barrier to reaction. The rapid decrease after the optimum temperature is explained by thermal denaturation and loss of three-dimensional shape (conformation) of the protein including its active site. A sketched graph of the expected relationship is a helpful inclusion (Figure 9.1).

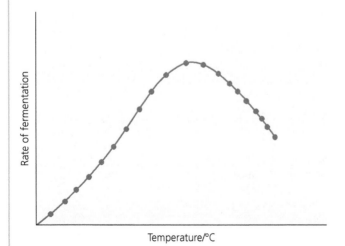

Figure 9.1 Sketch graph of rate of fermentation (in arbitrary units) versus temperature

Your hypothesis and subsequent investigation could be stimulated by an observation in nature. For example, you might have observed that the ground beneath certain species of tree, such as the black walnut (*Juglans nigra* L.), lacks vegetation. You might speculate that a soluble chemical washed from the leaves or tree roots could be responsible for the lack of plant growth. Research will lead you into the area of allelopathy, the release of chemicals by a plant that have an effect on other plants; this could form the basis of an Individual Investigation.

> **Examiner guidance**
>
> You should be aware of the falsification approach to the scientific method, which assumes that a scientific hypothesis must be testable and could potentially be shown to be false by experimental data. Ideally, your data will either support (not prove) your hypothesis or falsify (disprove) your hypothesis. You should be aware of the tentative nature of scientific knowledge.

Investigating relationships

If carrying out a statistically based investigation, there are three possible relationships:

- Is there a *correlation* between, for example, variable X (abiotic) and variable Y (biotic)?

- Is there a *difference* in mean/median measurement between, for example, habitat A and habitat B?

- Is there an *association* between categories A and B?

When selecting other variables to control or monitor, this is a good opportunity to use your biological knowledge to state what other variables might affect your dependent variable. What impact might they have on your dependent variable?

Minimizing errors

When designing an experiment take note of the following points:

- Ensure that the independent variable is the only major variable that is changed (manipulated).

- Include controls and comparisons to show it is only the independent variable that causes the measured biological or biochemical effect (if there is one).

- Where appropriate, select experimental subjects randomly to cancel out variation arising from biased selection (this is important in ecological investigations).

- Keep the number of replicates as high as possible, given the time constraints of the Individual Investigation.

- Ensure the same number of replicates is done for each value of the independent variable.

- Identify other factors that could affect the dependent variable and keep them constant (controlled variables, for example, temperature, pH, volumes of solutions, light intensity, time for reaction, etc.). Variables that cannot be controlled, such as those in ecological investigations (pages 71–2), need to be monitored.

Multifactorial investigations

Multifactorial experiments consider the interaction of two or more experimental treatments. This is a useful design for looking at the combination of effects. For example, the action of lipase on fats in milk (as measured by a colour change in an acid–base indicator) can include the addition of bile salts (an emulsifying agent) as an additional treatment.

In this example, the experimental treatments are: lipase alone, bile salts alone, lipase and bile salts, no lipase or bile salts. This is a four-treatment (including a control treatment) multifactorial experiment.

Multifactorial experimental designs allow a more complex range of research questions to be tested and can be used to investigate interactions that are likely to occur, such as: 'How does lipase affect fats in milk?'; 'How do bile salts affect fats in milk?'; 'Are the effects of bile salts and lipase on fats in milk independent, or is the effect of one influenced by the presence or absence of the other?'

Data tables

The following guidelines should be followed when presenting data in tables.

- All raw data should be presented in a single table with ruled lines and a border.

- Ideally, each table of biological results should show one relationship between the independent and dependent variables.

- Put the data for the independent variable in the first column; put the data for the dependent variable in columns to the right.

- Put processed data (for example, means, rates, reciprocals, logarithms and standard deviations, etc.) in columns to the far right.

- No calculations should be present in the table, only calculated values.

- Each column should be headed with a physical quantity, correct units and absolute uncertainty; the units should be separated from the physical quantity using a solidus (forward slash).

- No units should be in the body of the table, only in the column headings.

- Raw data should be recorded to a number of decimal places appropriate to the precision (sensitivity) of the apparatus or instrument.

- All raw data of the same type should be recorded to the same number of decimal places.

Risk assessment

Your IA report must contain a risk assessment where relevant (for example, this is not required for a simulation experiment). The three main parts of a risk assessment are:

- Hazard identification: identifying safety and health hazards associated with laboratory work or fieldwork.

- Risk evaluation: assessing the risks involved.

- Risk control: using risk control measures to eliminate the hazards or reduce the risks.

Detailed information about how to carry out risk assessments can be found on pages 37–9.

Additional hazards and risks associated with fieldwork include:

- Terrain, which refers to how the land lies. Variations in terrain might include uneven surfaces, flat areas, hills and steep gradients. It is important to select appropriate shoes.

- Weather conditions can change very quickly in the field. A weather forecast should be consulted before setting out, and appropriate clothing, footwear and supplies selected. In extreme weather, fieldwork might have to be postponed or abandoned.

- Areas where fieldwork is carried out can be isolated. It is essential that the school and parents who are not going into the field are aware of the route and the expected time of return.

- Tides can change very quickly. Tide tables should be consulted before setting out.

Environmental and ethical assessment

You need to consider the impact of your investigation on any organisms you are studying and the environment they live in, if relevant to your study. Bear in mind the IB ethical policy.

Exploration criterion checklist

Defining the problem and selecting variables

Descriptor	Complete
Research question	
You identify an appropriate topic and a general aim for investigation.	
You state a relevant and specific research question. research question is clear and sharply focused.	
Your research question refers to independent variable and dependent variable.	
Your research question sets the framework for the entire Individual Investigation and is consistently carried through.	
You state and describe the correct independent (or manipulated) variable, including units.	
You state and describe the correct dependent variable, including units.	
You predict, when appropriate, a quantitative relationship between the independent variable and the dependent variable(s).	
You state and describe the relevant controlled variables, including units, together with why and how they are controlled or monitored.	
Background information	
You give detailed relevant and appropriate biological background theory and information that enhances understanding of the investigation and puts it into a biological context.	
You include a hypothesis or biological model (quantitative or qualitative), where appropriate.	
You outline any assumptions or simplifications in any biological models or theories.	
You include a brief survey or summary of the biological literature that is referenced according to a stated referencing style.	
Safety, ethical and environmental issues	
The plan shows awareness of safety, ethical or environmental issues related to the methodology; for example, risk assessment, use of biological organisms/materials and issues related to storage and disposal of chemicals.	

Controlling/monitored variables

Descriptor	Complete
You include a description of apparatus, equipment and instrumentation (including range, sensitivity and absolute uncertainty in a single measurement).	
You include a description of quantity (mass/volume (with units)), concentration (with units) of solutions, physical state of solids, etc. A single reference in the method is acceptable.	
You give a clear, detailed and logical sequence of reproducible steps.	
You describe the rationale or justification of relevant steps in the method.	
You describe how your methodology minimizes random errors.	
You describe any calibration and checking for systematic errors.	
You include a cross-sectional and labelled diagram (photos, where appropriate) showing arrangement of non-standard apparatus. Correct names and terminology are used.	
You describe how and why controlled variables are to be held constant or otherwise monitored (if they cannot be controlled).	
You describe how the independent variable is varied and the values/range chosen for manipulation.	
You describe how the dependent variable is measured and how the processed variable(s) is/are deduced from the raw data.	
You outline planned controls (if appropriate).	
You explain choices with regard to the methodology, equipment or instrumentation selected and materials or substances.	
You have explored alternative methods and outlined why they are less suitable.	
Method complete and could be replicated.	
Equipment list provided.	

Planning and recording of data

Descriptor	Complete
You plan to collect a sufficient number of reliable and relevant raw data points over a wide data range.	
You plan to collect a suitable number of repeated and averaged readings.	
You plan to collect relevant qualitative data (observations).	
Your method takes into account and minimizes likely random and systematic errors in the raw data.	
You ensure that your data collection is relevant to the initial research question.	
You plan to collect sufficient data for any statistical analysis (if appropriate).	
You plan to collect raw data that record units and random uncertainties, as well as being recorded to an appropriate precision.	
You plan to record physical conditions, such as climate (for example, temperature; wind speed), if these affect the value of your dependent variable(s).	

10 Analysis

This criterion assesses the extent to which your report provides evidence that you have selected, recorded, processed and **interpreted** the data in ways that are relevant to the research question and can support a conclusion.

Mark	Descriptor
0	The student's report does not reach a standard described by the descriptors below.
1–2	The report includes **insufficient relevant** raw data to support a valid conclusion to the research question.
	Some **basic** data processing is carried out but is either too **inaccurate or too insufficient to lead to a valid** conclusion.
	The report shows evidence of little consideration of the impact of measurement uncertainty on the analysis.
	The processed data are incorrectly or insufficiently interpreted so that the conclusion is invalid or very incomplete.
3–4	The report includes relevant but incomplete quantitative and qualitative raw data that could support a simple or partially valid conclusion to the research question.
	Appropriate and sufficient data processing is carried out that could lead to a broadly valid conclusion but there are significant inaccuracies and inconsistencies in the processing.
	The report shows evidence of some consideration of the impact of measurement uncertainty on the analysis.
	The processed data are interpreted so that a broadly valid but incomplete or limited conclusion to the research question can be deduced.
5–6	The report includes sufficient relevant quantitative and qualitative raw data that could support a detailed and valid conclusion to the research question.
	Appropriate and sufficient data processing is carried out with the **accuracy** required to enable a conclusion to the research question to be drawn that is fully **consistent** with the experimental data.
	The report shows evidence of full and appropriate consideration of the impact of measurement uncertainty on the analysis.
	The processed data are correctly interpreted so that a completely valid and detailed conclusion to the research question can be deduced.

© IBO 2014

Table 10.1 Mark descriptors for the Analysis criterion

Recording and presentation of raw and processed data

The presentation of raw data needs to be accurate and qualitative observations included. Raw data must then be processed in some way: at its most basic, this will take the form of calculating means and standard deviations. You can use programs, such as Microsoft Excel, to analyse data (see pages 103–6).

Raw data – these are the data collected without any processing. They are just the values of each variable collected. They are often difficult to use for data analysis, and usually need to be processed in some way. Observations are another example of raw data.

Processed data – data that are ready for analysis. Processing can include merging or transforming data.

Numerical raw data must be presented in tables with appropriate headings and units (see page 122). The number of decimal places used for each variable must be consistent with the precision of the equipment used (for example, if a balance can measure to 2 decimal places, this is the precision that should be used to record the raw data). The same number of decimal places must be used consistently across all data for each variable.

When large amounts of data have been collected, using, for example, data logging, you should only present a representative sample of the raw data.

You need to ensure that your sample size is large enough for data processing:

- >30 is considered a large sample

- 15–30 a small sample

- 5–15 a very small sample

- <5 is usually considered too small a sample to calculate standard deviation or apply tests such as the *t*-test (see page 91).

Data processing

If carrying out a statistical test (pages 89–100), your investigation should have both a hypothesis and null hypothesis. For example, if carrying out a test to see if there is an association between two variables *x* and *y*:

- **Hypothesis** – There **is** a statistically significant association between variables *x* and *y*.

- **Null hypothesis** – There is **no** statistically significant association between variables *x* and *y*.

The concluding statement of your investigation report should cover the following points:

1 What is the result? State the outcome of your statistical test and whether it is larger, equal to or smaller than the critical value at your specific degrees of freedom at 5 % significance level.

2 What does this mean? Do you accept or reject the null hypothesis?

3 How confident are you? What is the probability that the result has happened by chance (this is set at 5 % in ecological studies, which means you can be 95 % certain that your results show a real association between the two variables).

4 Explain the biology.

Presenting data – graphing

One of the most effective ways to communicate the results of a scientific investigation is to create an effective visual representation (a graph) of the data that have been counted, measured and calculated. Often you can easily see patterns in a graph that might not be obvious in a data table. Graphs show trends in data clearly, and can be used to summarize complex relationships between independent and dependent variables. You need to select the correct graph to use for your data:

- ■ Categorical data – bar chart (± error bars/variation in data)
- ■ Continuous data – line graph.

Figure 10.1 can be used to select the correct graph for your data.

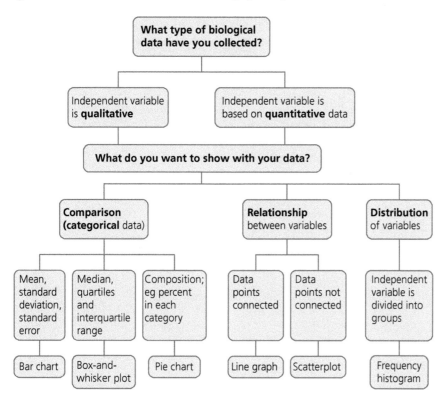

Figure 10.1 Flow chart for selecting the appropriate graph

Common mistake

Students sometimes confuse the use of bar charts, histograms and line graphs. The following should help you to distinguish between them:

Bar charts: The independent variable is a category, for example, blood type (A, B, AB or O), with the number of individuals in each category recorded on the *y*-axis:

- • Data in categories on the *x*-axis
- • Categories in any order
- • Bars do not touch

Histogram: The independent variable is not grouped into a category but into a range of numbers, for example, number of individuals at different ages in a population, where the *x*-axis divides population into groups of different age ranges, such as 10–19, 20–29, 30–39, etc., and frequency (number of individuals) in each group is recorded on the *y*-axis:

- • Data in ranges on the *x*-axis
- • Ranges placed in numerical order
- • Bars can touch

Line graph: Shows a trend in data:

- • Both variables show numerical (number) data
- • Data on *x*-axis (independent variable) are numbered
- • *x*-axis data are specific data points, not ranges
- • Points are placed in numerical order
- • A line can be drawn through the points, or a line of best fit added. The line should not go beyond the first or last points

Expert tip

It is suggested that you draw a graph of the results while the experiment is in progress. Any data not conforming to the trend can be identified and re-assessed, either by immediately recording the measurement again, or, if this is not possible due to the experimental conditions, by recording that measurement again before the equipment is dismantled.

Expert tip

Different types of graphs, charts and diagrams include: bar chart, pie chart, frequency histogram, timeline, line graph, scatter, kite, Venn, box-and-whisker plot, Gantt chart, frequency polygons and mindmap.

■ **ACTIVITY**

1 Find out about the following types of graph or other ways of representing data: histograms and bar charts, pie charts, box-and-whisker plots, triangular graphs, kite diagrams, and the ACFOR scale. **Summarize** how they are used, and give examples for each.

■ **Logarithmic scales**

Sometimes during an IB Biology investigation, the data collected are not easy to plot on a graph because of the very wide range of numbers involved. An example arises when measuring the growth rate of yeast using a counting chamber known as a hemocytometer. Rather than plotting the averaged raw data, it is preferable to calculate the base ten logarithm of the measurements (Figure 10.2).

This can be done with a calculator or Excel: the formula is =LOG10(cell reference). The logarithm of the number increases relatively slowly and a straight-line plot is generated. Alternatively, log-linear graph paper (Figure 10.3) can be used.

Common mistake

Graphs are sometimes reduced in size to ensure the report is the correct page length. Graphs should not be reduced to such a size that they become uninformative, simply to stay within the page limit.

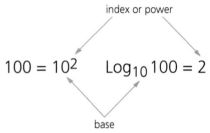

index or power

$$100 = 10^2 \qquad \text{Log}_{10}\, 100 = 2$$

base

Figure 10.2 Relationship of numbers with their logarithms

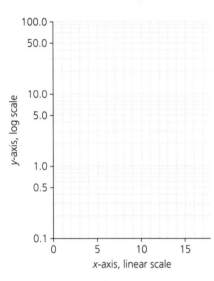

y-axis, log scale

x-axis, linear scale

Figure 10.3 A logarithmic scale on the *y*-axis with three cycles: 0.1 to 1.0, 1.0 to 10.0 and 10.0 to 100. The *x*-axis has a linear scale. Numbers can be plotted directly onto graph paper without calculating their $\log_{10}$ values.

Examiner guidance

- Graphing, even that of raw data, is part of processing, especially if it is used to derive values (for example, gradients for rates).
- Graphing of raw data when the graphing of processed data would be more appropriate is not incorrect but can be considered insufficient.
- Dot-to-dot plotting of data is acceptable for continuous data. Going further and placing a trend line on the data, especially if it is accompanied by error bars, is advisable. Calculating a correlation coefficient (*r*) can be a useful step in processing and interpretation (for example, by comparison with an accepted model).
- You need to obtain sufficient data so you can be confident when drawing a trend line on your graph. A trend line may be used to show how the limited data collected fits a given model (for example, pH optimum for an enzyme).
- Standard deviation or standard error can be useful assuming there are a sufficient number of replicates for calculation. Alternatively, range bars are acceptable for maximum-minimum values.
- The types of graph produced should be appropriate for the data being analysed.

Variation in data

In biology, the biggest issue relating to uncertainties is the variation in the biological material: this variation can be expressed as standard deviations (SD), standard error (SE), or the maximum-minimum range. In addition, measurement uncertainty can be calculated for apparatus used in the investigation (see below). Error bars showing variation should be used on graphs and their significance explained. You can show the variation in data on graphs by plotting SD, SE or minimum and maximum values above and below mean values.

■ Impact of measurement uncertainty

You are expected to appreciate the limitations of your instruments, and to present and discuss measurement uncertainties. The aim of a scientific investigation is to test a hypothesis and to record, as well as possible, the true value of a dependent variable (that is, to make the experiment accurate). The experimental procedure and apparatus will, however, result in measurements that vary from true values. These are known as 'uncertainties'. There are several different types of uncertainty:

Reading uncertainty: For analogue instruments (for example, a metre rule or an alcohol thermometer) the scale can only be read to within a certain fraction of the smallest scale division. This is usually taken to be half the smallest division. For digital instruments (for example, a pH meter and probe, or electronic balance) it is normally taken to be ±1 of the smallest change in reading.

Calibration uncertainty: This is a calibration provided by an instrument maker against approved standards; for example, for a metre rule it might be stated that the length of 1 metre is accurate to ± 0.5 mm. With increasing age and use this calibration might not be maintained.

Random uncertainty: If a particular procedure is repeated many times, the result might not be the same on every occasion, causing readings to vary in an unpredictable way from one measurement to the next. This could be because the equipment is set up slightly differently, or read slightly differently (for example, see parallax error, Chapter 1 page 11). These random differences will lead to a range of results which, assuming a normal distribution, can be statistically analysed to give a best estimate and an uncertainty for the measurement. The effect of random errors can be reduced by carrying out repeats and averaging precise results.

Systematic effects: These cause readings to vary from the true value by the same amount every time a measurement is made. Repeats *will not* compensate for systematic errors. Different apparatus or a different technique should be used and the results compared. Calibration or re-calibration of apparatus might solve systematic errors. These uncertainties differ from other uncertainties because they affect the results in the same direction. For example, if you assume a ruler's scale zero is at the end of the ruler, when it is in fact 1 mm from the end, then all the measurements made will be systematically 1 mm too small.

Zero errors: If a measuring system gives a false reading when the true value of a measured quantity is zero – that is, if equipment is not calibrated properly – this will cause a systematic error. For example, when using a balance, if the instrument is not set to zero before each measurement then the apparatus will give false readings.

see p. 11

> ### Expert tip
> The most common way of showing uncertainty is by giving a measured value ± uncertainty. For example, a measurement of 4.01 g ± 0.01 g means the experimenter is confident that the actual value for the quantity being measured lies between 4.00 g and 4.02 g.

> ### Expert tip
> Uncertainties may also be referred to as 'errors', for example, 'random error'. Error is the difference between the true value and the measured value, and uncertainty describes the range of values within which the true value is asserted to lie.

> ### Common mistake
> Error does not refer to a mistake; error refers to the value of the uncertainty.

> ### Examiner guidance
> You must recognize that all measured values have uncertainty and are not exact.

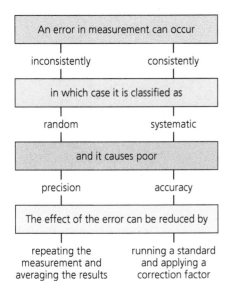

Figure 10.4 Concept chart contrasting random and systematic errors

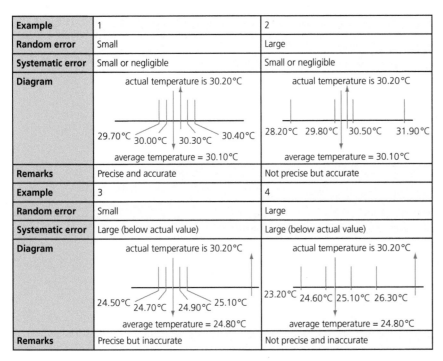

Figure 10.5 Examples of random and systematic error in temperature measurements

■ Multiple instances of measurements

Some methods of measuring involve the use of multiple instances to reduce the random uncertainty – for example, measuring the thickness of several leaves together, rather than a single leaf. The random uncertainty of each measurement is the random uncertainty of the whole measurement divided by the number of leaves. This method works because the percentage random uncertainty of the thickness of a single leaf is the same as the percentage random uncertainty for the thickness of multiple leaves.

Worked example

Thickness of 10 leaves: (4.10 ± 0.1) mm. Mean thickness of one leaf: (0.41 ± 0.01) mm.

◼ Combining random uncertainties (error propagation)

Where a sequence of experimental procedures, each with their own uncertainty, combine to contribute to overall uncertainty, this is known as error propagation.

Combination	Operation	Example
Adding or subtracting values	Add the absolute uncertainties	Mass of weighing bottle and sucrose = (18.54 ± 0.01) g Mass of weighing bottle = (13.32 ± 0.01) g Mass of sucrose = (5.22 ± 0.02) g (*weighing by difference*)
Multiplying values	Add the percentage uncertainties	Mass of water = (50.0 ± 0.1) g Temperature rise = (10.8 ± 0.1) °C (from burning dried food) Percentage random uncertainty in mass = 0.20 % Percentage random uncertainty in temperature = 0.93 % Enthalpy change = 2259 J (*from product of mass of water, specific heat capacity and temperature change*) Percentage random uncertainty in heat change = 1.13 % Absolute random uncertainty in heat change = ± 26 J (*Note – the random uncertainty in the specific heat capacity of water (4.184 J g^{-1} °C^{-1}) is taken to be zero*).
Dividing values	Add the percentage uncertainties	Mass of potassium iodide in solution = (100.0 ± 0.1) g Volume of potassium iodide solution = (250.0 ± 0.5) cm^3 Percentage random uncertainty in mass of potassium iodide = 0.10 % Percentage random uncertainty in volume of solution = 0.20 % Concentration of potassium iodide solution = 0.400 g cm^{-3} = 40.0 g dm^{-3} Percentage random uncertainty of concentration of potassium iodide = 0.30 % Absolute random uncertainty of concentration of potassium iodide = ± 0.0012 g cm^{-3} = ± 1.2 g dm^{-3}
Power rules	Multiply the percentage uncertainty by the power	Radius of leaf disc = (6.0 ± 0.1) cm Percentage random uncertainty in radius = ± 1.6 % Area of leaf disc = πr^2 = 113.1 cm^2 Percentage random uncertainty in area = ± 3.2 % Absolute random uncertainty = ± 3.6 cm^2

◼ Average of reading

Repeated measurements can lead to an average value for a calculated quantity. The final answer could be given to the propagated error of the component values in the average.

Worked example
Mean enzyme activity =
$$\frac{+100 \text{ mmol min}^{-1} (\pm 10\%) + 110 \text{ mmol min}^{-1} (\pm 10\%) + 108 \text{ mmol min}^{-1} (\pm 10\%)}{3}$$
Mean enzyme activity = 106 mmol min^{-1} (± 10 %)
Note: both the mean of the values and propagated error of the component values in the average are calculated by adding the values together and dividing by the number of sets of data (in this case, 3).

▓ Multiple readings

If more than one reading of a measurement is made, the random uncertainty increases with each reading. For example, 10.00 cm³ of hydrochloric acid is delivered from a 10 cm³ pipette (±0.10 cm³), repeated 3 times. The total volume delivered is 30.00 ± 0.30 cm³.

▓ Calculating percentage error for independent and dependent variables

In your analysis section, you need to, if appropriate, calculate uncertainties (percentage errors) for both independent and dependent variables (uncertainties cannot be calculated if data are categorical or are counts of organisms), so that you can discuss the impact of these uncertainties in your conclusion and evaluation.

To estimate the uncertainty of your data, take the smallest value in your data set and calculate the percentage error of this value. The percentage error will be largest at this value, and so if this is not significant, the error in larger values is likely also to be not significant:

(error ÷ smallest value) × 100

■ For example, if measuring leaf length, absolute error = 0.5 mm, so if smallest value is 40 mm, percentage error = (0.5/40) × 100 = 1.25 %

■ Any uncertainty <10 % is unlikely to be significant (although the significance of an error depends entirely on the degree of accuracy you are aiming for).

Table 10.2 shows absolute errors for some common apparatus used during ecological investigations.

Apparatus	Uncertainty (absolute error)
Oxygen probe	± 0.4 mg L⁻¹
Hygrometer	± 5 % rh
Anemometer	± 5 % m s⁻¹
Digital thermometer	± 0.1 °C
Digital balance	± 0.2 g
Light meter (Lux)	2,000 ± 0.5 20,000 ± 5 200,000 ± 50
pH probe	± 0.1
Flow meter	± 1.5 % m s⁻¹
Non-electronic apparatus	± 0.5 × smallest unit of measurement

Table 10.2 Absolute errors of ecological apparatus. Some errors are expressed as units and others as a percentage of the data value recorded.

Interpretation of processed data

You need to interpret your processed data correctly, for example, you should be able to deduce from a graph the correct graphical relationship between the independent and the processed dependent variable, and quote data to support your ideas. You should also analyse the results of any statistical tests you have carried out (see Chapter 6, pages 89–100), if appropriate. The interpretation of the data can be presented after each data set.

Expert tip

Certain plot shapes are associated with models that make it easier to suggest causal mechanisms. For example, a bell-shaped curve is associated with random samples and normal or Gaussian distributions; a concave increasingly upward curve is associated with exponentially increasing functions, such as the early stage of bacterial or yeast growth; an S-shaped curve is associated with a carrying capacity of the environment (a logistic curve); and a sine-wave-like curve is associated with a biological rhythm.

Analysis criterion checklist

▦ Recording raw data

Descriptor	Complete
You neatly record **all** raw data: qualitative data (observations) and quantitative (numerical) data necessary to support a conclusion to the research question.	
You present all raw numerical data clearly and correctly in tabulated manner (for example, independent variable on far left, then dependent variable followed by processed variable). SI units are typically used.	
The headings in your data tables have labels, units and uncertainties (absolute errors) once in the headings.	
You consistently record quantitative data taking into account the absolute uncertainty/error (correct number of decimal places).	
If using scientific notation, you quote the value and the error with the same exponent.	
You describe how the absolute uncertainties in the measurements were obtained, for example, half of the least count of the scale or manufacturer's tolerance.	
You mention reaction time for manual timings.	
You clearly indicate and highlight any anomalous data (and statistically justify their exclusion, if appropriate).	
Your processed data resolve the research question.	

▦ Processing raw data

Descriptor	Complete
You use the appropriate formulas and annotated mathematical equations to carry out calculations to process raw data (for example, averages, rates, percentage change, reciprocals or logarithms (to base 10 or e)).	
You include descriptive statistics, for example SD, range.	
You record results of calculations according to the rules of significant figures.	
You convert the appropriate averaged data into the correct graphical form: line graph, scatter graph, bar chart, etc.	
Where relevant you select appropriate processed average data to produce a straight-line graph (if data are continuous) (with line of best fit: graphically or computer generated).	
Your graphs include the equation and the R^2 value, where appropriate.	
You extract relevant quantity(ies) from the line graph, for example, gradient of a straight line, gradients along a curve, intercept (or rarely area) or perform interpolation or extrapolation.	
You process replicate data – finding the mean and using the variation between values to assign an appropriate uncertainty.	
Statistical test (if appropriate)	
Statistical test identified and its justification explained.	
Hypothesis and null hypothesis stated.	
Accurate processing of data for statistical test present.	
Result of statistical test clearly stated.	
Comparison of statistical test result to critical value present.	
Clear and correct statement as to whether the hypothesis is accepted or rejected.	

▦ Presenting processed data

Descriptor	Complete
You use scientific conventions in tables of processed data.	
You use accepted conventions for graphs, for example, correct graph size (large), appropriate range and scale, labelling, units, etc.	
Your graphs have the independent variable on the x-axis; dependent variable on the y-axis.	
You draw the line or curve of best fit correctly and clearly indicate the trend shown.	
You include error bars on line graphs, where appropriate and possible.	

■ Impact of uncertainty

Descriptor	Complete
You convert absolute uncertainties to percentage errors. Calculation of percentage error for at least one value of the independent variable and dependent variable (where appropriate).	
You present calculations in a clear, organized and separate manner.	
You present final values with the number of decimal places consistent with the uncertainty.	
You explain the impact of measurement uncertainty relating to the independent variable and dependent variable (where appropriate).	

■ Interpreting processed data

Descriptor	Complete
The processed data are correctly interpreted, for example, the correct graphical relationship between the independent and the processed dependent variable is deduced from a graph with data quoted to support it.	
Analysis of the correct statistical test is carried out, if appropriate.	

 Evaluation

This criterion assesses the extent to which your report provides evidence of evaluation of the investigation and the results with regard to the research question and the accepted scientific context.

Mark	Descriptor
0	The student's report does not reach a standard described by the descriptors below.
1–2	A conclusion is **outlined** which is not relevant to the research question or is not supported by the data presented. The conclusion makes superficial comparison to the accepted scientific context. Strengths and weaknesses of the investigation, such as limitations of the data and sources of error, are **outlined** but are restricted to an **account** of the **practical** or **procedural issues** faced. The student has **outlined** very few realistic and relevant suggestions for the improvement and extension of the investigation.
3–4	A conclusion is **described** which is relevant to the research question and supported by the data presented. A conclusion is **described** which makes some relevant comparison to the accepted scientific context. Strengths and weaknesses of the investigation, such as limitations of the data and sources of error, are **described** and provide evidence of some awareness of the **methodological issues** involved in establishing the conclusion. The student has **described** some realistic and relevant suggestions for the improvement and extension of the investigation.
5–6	A conclusion is **described and justified** which is relevant to the research question and supported by the data presented. A conclusion is correctly **described and justified** through relevant comparison to the accepted scientific context. Strengths and weaknesses of the investigation, such as limitations of the data and sources of error, are **discussed** and provide evidence of a clear understanding of the **methodological issues** involved in establishing the conclusion. The student has **discussed** realistic and relevant suggestions for the improvement and extension of the investigation.

© IBO 2014

Table 11.1 Mark descriptors for the Evaluation criterion

Common mistake

This is often the weakest criterion in a report. It is a difficult skill but sometimes candidates finish off the report in a hurry, leading to an inadequate evaluation.

Conclusion

In this criterion you need to state a conclusion based on the outcomes from your investigation. Conclusions need to be supported by the data and explanations included. You must refer back to your research question at this point in the report. A scientific context is needed for a full discussion.

The conclusion should focus on how the independent variable affects the dependent variable.

Expert tip

In a statistical analysis, if there is no significant result then discuss possible effects of monitored variables on the dependent variable.

Strengths and weaknesses of the investigation

The methodology must be evaluated thoroughly. You need to include the strengths as well as the weaknesses of the investigation. For example, a strength could be a small standard deviation in data for the dependent variable, indicating that measurements were precise. Weaknesses could include instances where other variables in an investigation were not sufficiently controlled, or if using an online simulation the software might not have accurately reproduced natural conditions (such as the natural selection simulation discussed on page 110). Do not restrict weaknesses to details of the practical, but include other aspects as well (for example, lack of secondary data or literature values for comparison).

Limitations of the data and sources of error

In this part of the report you can discuss uncertainties (percentage error) of dependent and independent variables (see pages 129–132). The impact of the limitations on the conclusion must be discussed. Your report needs to propose improvements, and these must be realistic and specific.

Improvements and extensions

You need to discuss how you could avoid the limitations you have outlined. For example, if a large range of pHs had been tested (pH 3–11) during an investigation of the effect of pH on the digestion of casein (a protein) by neutrase (a protease enzyme), a greater number of pHs, at greater resolution (sensitivity) could be tested either side of the suggested optimum pH (pH 6) for the enzyme (for example, pH 5–7 at 0.25 intervals).

You also need to suggest an extension to the work you have carried out: this must follow on logically from your investigation. Extensions should not just be 'more of the same' (for example, simply repeating the same experiment with a greater frequency of independent or dependent variable); instead, a significant extension is required, such as a different dependent variable, different species studied, and so on. For example, if measuring the change in mass of potato tissue due to osmosis (see Chapter 3, page 46), an extension could compare the gain in mass of potato tissue taken from the edge of the potato to that from its centre, in order to investigate how water moves through the tissue. If measuring the effect of soil moisture on leaf length of a plant species such as yarrow (*Achillea millefolium*) across an ecological **succession**, an extension could be the measurement of soil depth rather than moisture, or measurement of average plant height rather than the height of one specific species.

> **Expert tip**
>
> In an ecological study, focus your Evaluation on the limitations of independent and dependent variables, and sampling methodology, rather than monitored variables, and link specific improvements to them.

> **Expert tip**
>
> In your Evaluation, make sure you:
>
> - discuss the strengths of your investigation – these might be general or might refer to specific parts that worked well
> - discuss the reliability of the data
> - identify weaknesses in the method and materials
> - evaluate the relative impact of a weakness on the conclusion.

Evaluation criterion checklist

Concluding

Descriptor	Complete
Your conclusion is clearly stated, which is in accordance with the statistical test result, scientific explanation and impact of measurement uncertainties.	
You include a detailed conclusion (for example, trend between independent and dependent variables) citing numerical values relevant to the research question.	
Your conclusion is supported by the raw and processed data (typically graphs or charts) and the observations.	
Statement of the result of the statistical test present, including significance, if relevant.	
You include a conclusion that is described, justified and compared to the relevant biological literature (if available).	
You identify and comment on any anomalous data.	
Your conclusion is based solely on your results and uses tentative words such as 'indicate', 'suggest', 'appear to suggest' and 'support'; not 'prove'.	
You include a valid conclusion relevant to the research question and that is within the limits of random uncertainties; the impact of measurement uncertainties has been considered.	
You justify your results with reference to relevant biological models, theories and principles.	
You discuss limitations to your results.	

Evaluating procedures: strengths and weaknesses

Descriptor	Complete
You describe assumptions that were made which have affected the accuracy of the results; for example, all organisms behave the same under the experimental conditions; negligible mortality during the experimental period; oxygen-enriched bubble size the same diameter; enzyme kinetic models are based on steady state kinetics; a negligible change in the concentration of the enzyme–substrate complex during the course of the reaction.	
You identify limitations of the data and sources of error associated with the method.	
You calculate the total experimental error and compare to the literature value (if available).	
You identify systematic errors and their directional effect on the experimental result (increase or decrease).	
You discuss any limitations of the method, for example, limited data range, limited instrument sensitivity, and sample size.	

Improving and extending the investigation

Descriptor	Complete
Specific modifications/improvements relating to the limitations and errors stated above are given.	
You suggest appropriate modifications in the steps taken to improve the accuracy, precision and reliability of the results (by reducing random and systematic errors) or better control/monitoring of controlled variables.	
You suggest a reasonable alternative method or different instrumentation to obtain the same experimental data, or more accurate data.	
You discuss clearly how the suggested improvements or modifications would improve the reliability, precision and accuracy of the results.	
You propose realistic and relevant extensions to the study, for example, new data processing/data presentation, choice of new independent variable.	
Your suggested improvements are focused on the existing research question.	
Your extensions are focused on a new research question.	

12 Communication

This criterion assesses whether your investigation is presented and reported in a way that supports effective communication of the focus, process and outcomes.

Your report should be word-processed, page numbered and any equations may be embedded in the text of your report and formatted using the 'Equation Editor' tool in your word processor (Figure 12.1)

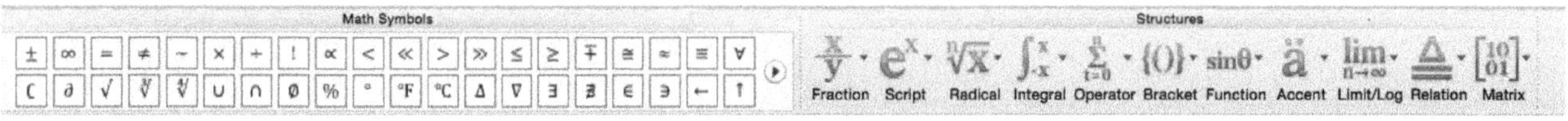

Figure 12.1 Equation Editor tool bar in Microsoft Word running on an i-Mac

Your report should be 6–12 pages in length, including any index or title page.

Mark	Descriptor
0	The student's report does not reach a standard described by the descriptors below.
1–2	**The presentation of the investigation is unclear, making it difficult to understand the focus, process and outcomes.** The report is not well-structured and is unclear: the necessary information on focus, process and outcomes is missing or is presented in an incoherent or disorganized way. The understanding of the focus, process and outcomes of the investigation is obscured by the presence of inappropriate or irrelevant information. There are many errors in the use of subject-specific terminology and conventions.*
3–4	**The presentation of the investigation is clear. Any errors do not hamper understanding of the focus, process and outcomes.** The report is well-structured and clear: the necessary information on focus, process and outcomes is present and presented in a coherent way. The report is relevant and concise, thereby facilitating a ready understanding of the focus, process and outcomes of the investigation. The use of subject-specific terminology and conventions is appropriate and correct. Any errors do not hamper understanding.

* For example, incorrect/missing labelling of graphs, tables, images; use of units, decimal places. For issues of referencing and citations refer to the academic honesty section.
© IBO 2014
Table 12.1 Mark descriptors for the Communication criterion

Structure and clarity

The presentation must be coherent and relevant to the focus (the research question and the process (the methodology)) and the outcomes (Results and Conclusion). It should resemble a research or review paper and there should be headings and sub-headings to give a logical sequence to the Individual Investigation report. Diagrams and digital images should be used to enhance understanding, with suitable referencing. There should be a logical flow to your report allowing the IB Examiner to understand your thought processes throughout the report. The description of your methodology should be detailed enough for the experiments to be reproducible, but simplistic, well-known and assumed aspects of your method need not be made explicit. No appendix should be included – it will not be read and will reduce your available word count.

Relevance and conciseness

Your written work, including the background information, should remain closely connected to and relevant to the research question and topic of your Individual Investigation. It should be easy for the IB Examiner to follow the development of your ideas and thoughts from the beginning to the end of your report. It should be concise with no unnecessary or repetitive (redundant) information. Your report should be 6–12 pages in length, including any index or title page. There are no automatic penalties for a report that is slightly longer, as long as it remains relevant and concise throughout.

Full calculations for processing of all numerical data and all error propagation are not expected – selected examples will be sufficient and free up more space for your conclusion and evaluation.

Make sure that tables of raw data are not repetitive, and use only one if this is the best way to present data. Similarly, avoid repetitive data tables when one would suffice. If you have drawn several graphs, it might be appropriate to combine them rather than display multiple graphs – this will not only save space but also allow you to draw comparisons between data.

If you have produced a graph from data logging which is then used to derive a value such as a rate, one example can be presented to explain the processing and then the rates derived can be organized in a table.

> **Expert tip**
>
> The Communication statement 'The report is relevant and concise thereby facilitating a ready understanding of the focus, process and outcomes of the investigation' is more likely to be met by a report of about 6 to 12 pages.

> **Common mistake**
>
> The following are often included but are not necessary: using whole pages for titles or contents; presenting blank data tables at the end of the method section.

Terminology and conventions

Pay close attention to the use of biological terminology (especially IB Biology terminology), scientific conventions (for example, letters in mathematical equations are italicized), units, significant figures, random uncertainties and nomenclature (binomial species names should be underlined or italicized).

> **Common mistake**
>
> The format of scientific names is sometimes incorrect, for example, having a capital letter for species name rather than the correct lower case letter; species names need to be presented in italics or underlined; for example, *Echium vulgare* or <u>Echium vulgare</u> not Echium vulgare or Echium Vulgare.

Examiner guidance

Remember that in any mathematical equation the units must balance on the two sides. For example, $0.08 \text{ cm}^3/2\text{ s} = 0.04 \text{ cm}^3 \text{ s}^{-1}$ and $4.0 \text{ cm} \times 4.0 \text{ cm} = 16.0 \text{ cm}^2$. It is good practice to include all units in calculations, or at least in one sample calculation.

Your report needs to reference material via footnotes, endnotes or in-text citations (see below). A full bibliography should be included at the end of the report.

Metric units should be included throughout the report. The numbers of decimal places should be consistently applied and correspond to the precision of the data. Measurement uncertainties should be included.

SI units should typically be used, with the exceptions of mass (grams) and volume (cubic decimetres and cubic centimetres).

All tables, graphs and equations should be introduced by a sentence of explanation. They should also have an explanatory label. The labels should be applied using the same formatting and numbered sequentially throughout the report, for example, equation (1).

■ SI units

You are normally expected to use the following units for recording measurements and in associated calculations during the course of the practical work carried out to support the IB Biology programme. Recorded measurements should always include the relevant units (and absolute error).

See also pages 4–5 for information about correct units to use in your report.

Quantity	Unit
Amount	Moles (not volume or mass)
Concentration	Mole per cubic decimetre, mol dm^{-3} or gram per cubic decimetre, g dm^{-3}. Parts per million can also be used (ppm)
Energy	Joule (J)
Specific heat capacity	Joule per gram per degree Celsius (J g^{-1} °C^{-1})
Mass	Gram (g)
pH; absorbance	No units (because they are logarithmic functions)
Rate of reaction	Moles per cubic decimetre per second, mol dm^{-3} s^{-1}. If gases are being measured then units of rate may be g s^{-1} or dm^3 s^{-1} or cm^3 s^{-1}. If a comparison of average rates is sufficient then use reciprocal of seconds (s^{-1})
Temperature	Standard thermometers measure temperature in degrees Celsius (°C)
Time	Seconds (s) (unless time intervals are long)
Volume	Cubic centimetres (cm^3) or cubic decimetres (dm^3). Measurements using laboratory apparatus will commonly be in cm^3, while concentrations are expressed in terms of dm^3

Table 12.2 Quantities and their associated units

■ Chemical formulas and names

Use subscripts, superscripts, parentheses and symbols with state symbols appropriately in chemical formulas, for example, Ca^{2+}(aq), KCl(s), Na_2CO_3(s), HCO_3^-(aq) and H_2O(l). Do not condense the formula, for example, use C_2H_5OH, rather than C_2H_6O to represent ethanol since the second formula also represents methoxymethane (CH_3-O-CH_3).

IUPAC names should be stated in brackets the first time the trivial name is used, for example, alcohol (ethanol), alanine (2-aminopropanoic acid), lime water (calcium hydroxide solution), copper sulfate (copper(II) sulfate), malic acid (2-hydroxybutanedioic acid) and citric acid (2-hydroxypropane-1,2,3-tricarboxylic acid).

However, complex molecules are usually referred to by trivial names, for example, starch, beta-carotene, human or bovine insulin, chlorophyll *a*, and beta-D-glucose.

Chemical names should not be capitalized unless named after a person, for example, Benedict's reagent or Knop's solution, or an abbreviation, for example, DCPIP, ATP, ADP, NAD^+ and HEPES.

Referencing

Referencing is an internationally or nationally standardized method of acknowledging the sources of information you have consulted when writing your Individual Investigation report. All words, paragraphs, quotes, figures, tables, digital images, theories, scientific ideas and facts originating from another source and used in your Individual Investigation report must be referenced (that is, acknowledged). Referencing is done for the following reasons:

■ to avoid plagiarism

■ so that your teacher can verify quotations

■ so that your teacher can follow up on your thinking by consulting the source you accessed.

There are many ways to acknowledge sources of information, for example, MLA (Modern Language Association), APA (American Psychological Association), Chicago, ACS (American Chemical Society), and Turabian: none is mandated by the IBO, though your school might dictate a referencing style that you must adopt. APA tends to be used by Psychology and the Sciences. Make yourself familiar with your school's requirements for referencing and citation. Be consistent and familiarize yourself with the format and terms that your school or IB Biology teacher expects you to use.

For example, the ACS style (shown below) has a citation consisting of two parts: the *in-text citation*, which provides brief identifying information within the text, and the *reference list*, a list of sources that provides full bibliographic information. In the ACS style the in-text citations can be referenced by superscript numbers, italic numbers or author name and year of publication.

Book

Davis, A.J.; Clegg, C.J. *Biology Study and Revision Guide for the IB Diploma;* Hodder Education, 2017.

Journal article

Dean, R.L. 'Treasure the exceptions': an investigation into why heat-treated, isolated chloroplasts fail to respond to uncoupling reagents; *J. Lab. Chem. Educ.,* 2014, Volume 2, No 4, 64–72.

Website

Perritano, J. Scientists discover caterpillar that actually eats plastic; https://animals.howstuffworks.com/insects/scientists-discover-caterpillar-that-actually-eats-plastic.htm (accessed May 4, 2017).

■ Assessing sources

- ▦ Can you identify the author's name?

- ▦ Can you determine what qualifications or titles he/she has?

- ▦ Do you know who employs the author, such as a university or company?

- ▦ Is this a primary source (original research paper) or secondary source (for example, a review article)?

- ▦ Is the content original or derived from other sources?

■ Evaluating information

It is important that you check the validity of the sources you are using. Do not assume that information is correct. The following checks can be made to ensure that the sources you use for your Individual Investigation provide you with accurate information:

- ▦ Have you checked a range of sources?

- ▦ Is the information supported by relevant literature citations?

- ▦ Is the information taken from a scientifically credible source (for example, a peer-reviewed scientific journal)?

- ▦ Is the age of the source likely to be important regarding the scientific accuracy of the information?

- ▦ Is the information scientific fact, opinion or speculation?

- ▦ Have you checked for any mistakes or inconsistencies in the arguments?

- ▦ Have the errors associated with any measurements been taken into account?

- ▦ Have the data been analysed (if appropriate) using relevant statistics?

- ▦ Are the data in graphs displayed fairly?

Expert tip

References are required in both background information (Exploration) and conclusions (Evaluation). Have other people done similar work to yours that you can refer to? Does your scientific information need referencing?

Expert tip

Scientific papers submitted to peer-reviewed journals, such as *Nature*, are carefully scrutinized by experts in the field ('peers') to ensure that the arguments, results and analyses presented are legitimate and worthy of publication. Information from such sources can be trusted as being scientifically valid.

Report format

There is no particular structure that you must follow for your Individual Investigation report, though it should resemble a research paper (without the abstract). If your school does not suggest a format, then you could use the following headings, or simply use the assessment criteria:

- General title or aim
- Background information
- Personal engagement
- Biological theory and hypothesis (if appropriate)
- Research question
- Risk assessment
- Planning and preliminary work
- Variables
- Methodology
- Statistics
- Controls
- Raw data
- Processed data (including graphs)
- Impact of measurement uncertainty
- Conclusion
- Evaluation
- Random and systematic errors
- Limitations, weaknesses and improvements
- Evaluation of secondary sources
- Future investigations
- Bibliography of references

Use Figure 12.2 to check that you have covered all aspects of each criterion.

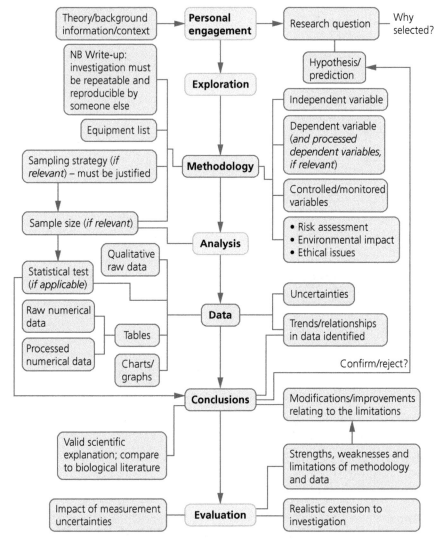

Figure 12.2 Biology Individual Investigation map

Expert tip

Carefully check your Individual Investigation report:

- The report has a structure which is clear, well organized, and easy to follow, for example, sections are clearly and helpfully headed.
- The report is logical and coherent.
- The report makes consistent use of appropriate terminology.
- The report is concise (between 6 and 12 pages).
- A sensible stance has been taken regarding font size and margin width, to ensure that good communication skills are demonstrated.
- Conventions have been followed in the presentation of tables, charts and graphs.
- Published works have been referenced correctly, with a suitable bibliography/reference list included.

Expert tip

Get a friend to read through and check your report. Do the ideas flow logically; is the use of independent and dependent variables consistent (in research questions, summary of variables), etc.?

Using tenses in scientific writing

When you write your report, you need to choose which tense to use.

Introduction

This is usually presented in the present tense, for example, *transect studies provide important information for ecological investigations*. When you are using the present tense you are indicating to the examiner that you believe the research findings are true and relevant, even though they were carried out in the past.

Methodology

It is usual practice (but not an IB requirement) to use the simple past tense with the passive voice to describe your experiments. For example, *the casein solution was added to the neutrase enzyme*.

Results

Results are usually described using the past tense. For example, *five samples of betalain solution, collected from beetroot immersed in distilled water at a specific temperature, were analysed with a colorimeter so that the level of absorbance could be recorded*.

Further discussion on this topic and additional examples can be found on the following website: http://services.unimelb.edu.au/__data/assets/pdf_file/0009/471294/Using_tenses_in_scientific_writing_Update_051112.pdf

Academic honesty

You need to ensure that the Individual Investigation report you submit for your Internal Assessment (IA) is your own work. At the end of the course you will need to sign a declaration to that effect. Throughout the course (and not just in Internal Assessment), your IB Biology teacher is required to ensure that any submitted work is your own.

When in doubt, your IB Biology teacher will check the authenticity of your work by:

- discussing it with you
- asking you to explain the method and to summarize the results and analysis
- asking you to repeat an investigation
- using software (such as turnitin.com) to check for plagiarism.

If you use any experimental method from another source then you need to ensure that you acknowledge it. This also applies if you use diagrams, tables, graphs, literature values from the Internet or reference books. Each lab report should have a bibliography acknowledging the sources used.

Plagiarism

Plagiarism is defined (by the IBO) as the representation, intentionally or unwittingly, of the ideas, words or work of another person without proper, clear and explicit acknowledgment.

Academic honesty and integrity are consistent with the IB learner profile (Introduction, page xiv), where learners strive to be principled. The IB upholds principles of academic honesty, which are seen as a set of values and skills that promote personal integrity and good practice in teaching, learning and assessment.

Examples of plagiarism include:

- copying the work of another IB Biology student (past or present) and passing it off as your own work
- using 'essay-writing' services, such as those available online or from a tuition centre

Figure 12.3 Never 'copy and paste' information – read and rewrite (or redraw if a figure) in your own words

Expert tip

The following suggestions will help you to avoid plagiarism:

- Make sure the work (results and theory) you present in your Individual Investigation report is always your own.
- Never 'copy and paste' from websites or Word or pdf files downloaded from the Internet.
- Place appropriate citations in your report, where required or appropriate.
- Show clearly where you are quoting directly from a source.

- copying text or images from a source (book chapter, journal article, or website, for example) and using this within your report without acknowledgment

- quoting others' words without indicating who wrote them or said them (personal communication)

- copying scientific ideas and concepts from a source without acknowledgment, even if you paraphrase them.

Communication criterion checklist

Descriptor	Complete
Structure of report	
Your report is well-structured, coherent and clear following the style and conventions of a scientific research or review paper.	
Your report is split into appropriate sections (with headings and titles) which could come under the broad headings of focus, process and outcomes, the IB criteria or other headings.	
Relevance and conciseness	
Your report only includes relevant information, is concise and 6–12 pages in length.	
Your report does not contain any errors, contradictions or false scientific statements or false assertions.	
Subject-specific terminology and conventions	
You have used appropriate scientific terminology throughout.	
Your graphs, tables and images are fully titled and referenced and presented according to conventions in scientific papers.	
Your mathematical equations are in italics and clearly explained and justified/derived (where appropriate) with units.	
You have followed the rules relating to significant figures. The number of decimal places is correct in data tables and calculations.	
You have defined non-syllabus terms and abbreviations.	
Your report includes a cross-referenced bibliography with in-text referencing according to a particular referencing style.	

Introduction

The **behaviour** of animals enables them to survive, to seek out favourable environments and to reproduce. Behaviour is based on feedback, using the control and coordination machinery of the body (sense organs, nervous system and the effector organs).

Behaviour is said to be either **innate** (instinctive, that is automatically triggered in certain circumstances) or **learned**. Learning occurs when experiences are retained and used to modify behaviour on future occasions.

The study of behaviour may provide a suitable topic for an Individual Investigation, provided the IB animal experimentation policy is complied with. Note that animal behaviour is far more complex and diverse than plant behaviour and hence difficult to investigate and explain with high validity. In particular, *'Experiments involving animals must be based on observing and measuring aspects of natural animal behaviour. Any experimentation should not result in any pain or undue stress on any animal (vertebrate or invertebrate) or compromise its health in any way. Therefore experiments that administer drugs or medicines or manipulate the environment or diet beyond that easily tolerated by the animal are unacceptable. Experiments resulting in the death of any animal are unacceptable'.* © IBO 2009

Investigations involving animal behaviour explore ways in which animals adapt themselves to changes in their environment though specific behavioural responses. Animal behaviour can be studied in the field, under natural conditions (this area of biology is known as **ethology**) or in the lab. Field studies are complex and are best carried out as a long-term investigation. Similarly, investigations of learned behaviour would be difficult to carry out over a limited timeframe. For short-term, controlled investigations, laboratory-based studies are recommended, involving innate behaviour. Invertebrates make ideal organisms to study, because they can be easily removed from the wild and studied under laboratory conditions. Orientation behaviour (see below) is a suitable area to investigate, as it will produce reproducible, controlled results over a relatively short period of time.

■ Orientation behaviour

Orientation behaviour involves coordinated movements that occur in response to an external stimulus.

Woodlice are an ideal organism for investigating orientation behaviour. Woodlice belong to a class of the arthropod phylum (animals with a hard external skeleton covering the segmented body, typically with a pair of jointed limbs per segment) known as crustaceans. Crustaceans are predominantly aquatic arthropods, and include the crabs and lobsters.

Woodlice are nocturnal and are typically discovered under logs and stones, in bark crevices, and among dead leaves and rotting plant material. They live on land but have not fully adapted to terrestrial conditions: their exoskeletons are not completely waterproof and almost all species of woodlice are confined to damp or humid places.

The environment provides conditions that are favourable to non-vertebrates such as the woodlouse, but is also the source of threats and dangers. Such threats may be both abiotic (such as excess heat, drought and danger of desiccation) and biotic (such as attack by predators). In woodlice, orientation behaviour favours their survival and reproduction (Figure A.1).

Key definitions

Behaviour – the way in which organisms respond to the environment and to other members of the same species.

Innate behaviour – behaviour that is inherited from parents and so develops independently of the environmental context. Innate behaviour includes behaviour that is due to a reflex action.

Learned behaviour – behaviour that develops as a result of experience.

Ethology – the study of animal behaviour in natural conditions.

Expert tip

- Orientation behaviour in non-vertebrate animals may improve the individual's chances of locating favourable conditions and remaining there, avoiding dangers and thus improving the chance of survival and reproduction.

- Woodlice cannot survive in dry conditions – orientation behaviour enables them to select habitats that are damp and dark, which affects their chances of survival and reproduction.

Common species

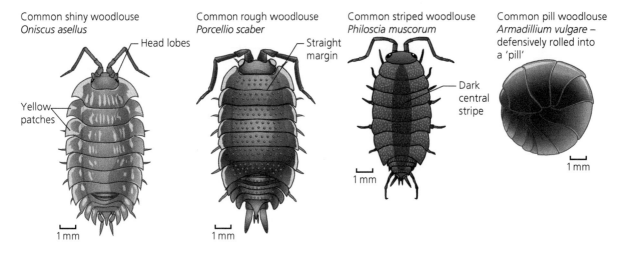

Common shiny woodlouse
Oniscus asellus
— Head lobes
Yellow patches
1 mm

Common rough woodlouse
Porcellio scaber
— Straight margin
1 mm

Common striped woodlouse
Philoscia muscorum
— Dark central stripe
1 mm

Common pill woodlouse
Armadillium vulgare – defensively rolled into a 'pill'
1 mm

Marked length of garden wall where woodlice are observed to browse and scavenge at particular times

In the hours of daylight the woodlice were found along the base of the wall, under stones, logs and rotting leaves.

Number of woodlice moving on the surface of the wall on nine occasions spaced out between dusk and dawn (study undertaken on a warm, humid night)

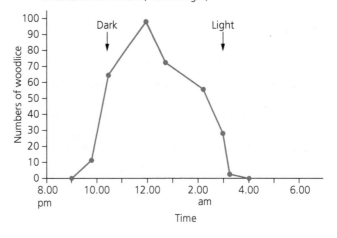

Figure A.1 Orientation behaviour of woodlice

■ **ACTIVITY**

1 **Examine** the pattern of orientation behaviour in woodlice, illustrated by the graph in Figure A.1. **Evaluate** the extent to which the data on woodlouse activity (Figure A.1) during the 24-hour period directly favours the survival and reproduction potential of the individual.

■ Choice chambers

Choice chambers are often used to study orientation behaviour. A choice chamber allows conditions to be changed in different parts of a container while keeping the animals under investigation out of direct contact with any chemicals used. Any large dish can be used, divided into segments with fine nylon mesh acting as a platform. A stoppered hole in the centre of a cover allows organisms to be introduced to the chamber. Anhydrous calcium chloride can be used to create dry conditions (placed under the mesh or muslin platform so that the chemical does not come into contact with the animals) and damp cotton wool can be used to create humid conditions in another part of the chamber. Small choice chambers can be made with Petri dishes (Figure A.2).

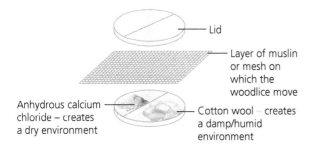

Lid

Layer of muslin
or mesh on
which the
woodlice move

Anhydrous calcium
chloride – creates
a dry environment

Cotton wool – creates
a damp/humid
environment

Figure A.2 Creating a choice chamber using a Petri dish

Ideas for investigations

A research question may ask whether 'woodlice show a preference for damp
areas in the choice chamber'. The null hypothesis would state that woodlice
will be distributed 50:50 between damp and dry areas. Woodlice are placed
in a choice chamber and, after a fixed amount of time, numbers of woodlice
in each part of the choice chamber are counted. The distributions expected
by the null hypothesis and those actually found in the choice chamber are
compared and statistically analysed using the chi-squared test (see Chapter 6),
which determines whether the final result significantly varies from the expected
distribution.

Alternatively, light and dark areas can be created in the chamber to test the
hypothesis that woodlice prefer dark conditions to light.

Orientation behaviour can be categorized as either *kinesis* or *taxis*. In kinesis, an
animal's body is not oriented in relation to a sensory stimulus, such as humidity, but
rather the stimulus causes a change in speed or direction of movement. Kinesis is
therefore a change in the rate of activity in response to an environmental stimulus.
In contrast, taxis is a change in movement in response to an environmental
stimulus, either towards (positive) or away (negative) from the stimulus.

A woodlouse placed in a dry container will move at a faster rate than one placed
in a humid environment – this is an example of kinesis. The faster movement
will give the woodlouse a greater chance (in the wild) of finding a humid, damp
environment.

Euglena is a protoctistan which contains chlorophyll and can photosynthesize, and
so requires light energy. If *Euglena* are placed in a Petri dish covered by aluminium
foil, excluding a few small exposed sections, and illuminated by a light source,
they will migrate towards the exposed sections – this is an example of taxis (the
Euglena display positive phototaxis).

Ideas for investigations

Rather than exploring taxis
responses in woodlice, as described
in the Investigation box above,
you could study kinesis behaviour
(that is, how stimuli affect rate of
movement).

Expert tip

In order to distinguish between
kinesis and taxis behaviour, a choice
chamber can be modified by having a
walk-way inserted that intersects the
chamber at the mid-line: the choices
made by woodlice at this point will
enable you to evaluate the possibility
of a taxis mechanism.

Ideas for investigations

Turn alternation is the other interesting woodlice behaviour that works well
in the laboratory. Woodlice will counter-turn where possible if they have been
forced to turn in a channel. This allows aspects of memory to be investigated –
for how long and how accurately do woodlice remember the forced turn?

Design of an animal behaviour study

The following questions and points will help guide the planning of an animal
behaviour investigation.

- Will the study describe a phenomenon, for example, kinesis, or test hypotheses?
  (*If the latter, you need to clearly define the behaviour (for example, innate
  behaviour) and identify the independent and dependent variables, and the control
  groups, if required.*)

- How will the variables be measured?

- What species of animal will be studied?

- What is the level of investigation (individual or group)?

- Will the study be performed in a biology laboratory or field setting? (The former is recommended for an IA.)

- Consider laboratory concerns, such as animal procurement (captured versus bred), housing and care of the animals, experimental conditions (humidity, light intensity and so on) and experimental techniques (such as choice chambers for studying orientation behaviour).

- If studying ethology, consider field concerns, such as how to make sure your investigation is minimally invasive, how to find, identify and track the animals and the data collection and recording tools you will use (for example, digital video footage, tape-recorder and data loggers).

- Consider legal and ethical concerns.

- Impact the smallest number of animals possible.

- Detect and reduce animal stress and pain.

- Evaluate the potential gain in biological knowledge against the consequences to the individual animals and populations.

- Use the appropriate control groups, test sequences, and statistical analysis.

Microbiology practical skills

Note: A full risk assessment must be carried out and teacher approval given before embarking on any practical microbiological investigation. The manufacturer's operating instructions for instruments, techniques and equipment or published papers with protocols must be consulted.

Aseptic techniques must be used to prevent contamination from pathogens. A resource on aseptic methods can be found here: www.nuffieldfoundation.org/practical-biology/aseptic-techniques.

> **Key definition**
> **Aseptic techniques** – procedures to prevent contamination from pathogens. These can include use of disinfectants and using a flame to kill micro-organisms.

Microbiological techniques

Micro-organisms, such as bacteria, are cultured on solid media, which are usually prepared using sterile agar, and in liquid media known as broths. These media contain an energy source, carbon source (for example, glucose), nitrogen source (amino acids), mineral salts and growth factors (for example, folic acid), all in aqueous solution.

To prepare agar plates, agar powder is used to make a dilute solution (1 % or 2 %) and boiled and left to cool to 42–45 °C (Figure B.1). Nutrients may be added unless a ready-made agar is used. While still hot and liquid the agar is poured into Petri dishes.

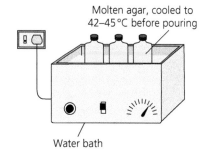

Molten agar, cooled to 42–45 °C before pouring

Water bath

■ ACTIVITY

1   Distinguish between sterilization and disinfection and find out how they are achieved in the laboratory.

Micro-organisms are usually cultured in pure culture, that is, without other organisms. One approach is by making a streak plate (Figure B.2). The aim of this aseptic technique is to obtain single isolated pure colonies of bacteria or yeast.

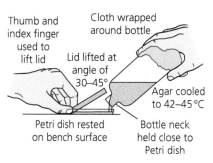

Thumb and index finger used to lift lid

Cloth wrapped around bottle

Lid lifted at angle of 30–45°

Agar cooled to 42–45 °C

Petri dish rested on bench surface

Bottle neck held close to Petri dish

Figure B.1 Pouring an agar plate

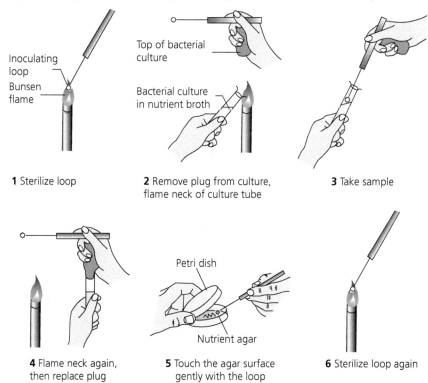

Inoculating loop
Bunsen flame

1 Sterilize loop

Top of bacterial culture

Bacterial culture in nutrient broth

2 Remove plug from culture, flame neck of culture tube

3 Take sample

4 Flame neck again, then replace plug

Petri dish

Nutrient agar

5 Touch the agar surface gently with the loop

6 Sterilize loop again

Figure B.2 Culturing bacteria on a streak plate: the Bunsen burner creates an updraft of air carrying away any bacterial or fungal spores from the open agar plate

■ ACTIVITY

2   Find out about the pour plate technique and its advantage over the streak plate technique.

Measuring micro-organism population growth

◼ Direct counting

Samples are removed and examined in a counting chamber, which is a thick glass microscope slide with a rectangular depression that creates a chamber. Counting chambers are used to determine the number of micro-organisms per volume unit of a liquid: the number of organisms is visually counted under a microscope. It may be necessary to dilute the sample with water. The disadvantage of this approach is that it gives total counts (dead and living yeast or bacteria).

◼ Viable counting

Samples are diluted and spread onto agar and colonies of yeast or bacteria develop. If the sample has been sufficiently diluted then the number of colonies can be counted.

◼ Turbidimetry

Samples are placed into a colorimeter (see Chapter 2, pages 32–3) and readings of absorbance taken. The greater the number of cells the more light is blocked, meaning less light is transmitted. This method does not distinguish between living and dead cells. Colorimeter readings can be converted to population numbers that use samples with known populations.

Testing for antibiotic sensitivity

Disc diffusion testing is a test of the sensitivity of bacteria to antibiotics. It uses paper discs soaked in antibiotic solutions to test the extent to which bacteria are affected by those antibiotics. Discs containing antibiotics are placed on an agar plate where bacteria have been placed, and the plate is left to incubate for 24 hours. The antibiotic diffuses out from the filter paper through the agar forming an invisible ring (Figure B.3).

If an antibiotic prevents the bacteria from growing or kills the bacteria, there will be an area around the disc where the bacteria have not grown enough to be visible, known as the zone of inhibition. The size of this zone depends on how effective the antibiotic is at stopping the growth of the bacterium. A stronger antibiotic will create a larger zone, because a lower concentration of the antibiotic is enough to stop growth.

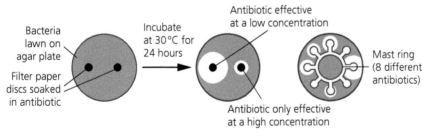

Figure B.3 Using antibiotic discs to determine the effectiveness of antibiotics

■ **ACTIVITY**

3 Figure B.4 shows the results of an experiment investigating inhibition of bacterial growth by bactericides on sterile bacterial cultures. **Analyse** the results and then **explain** your conclusions.

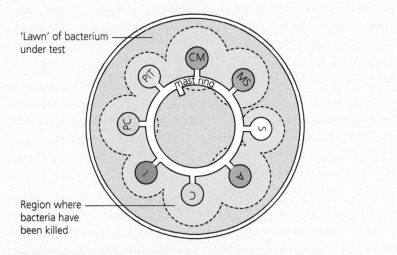

Figure B.4 Investigating sensitivity to bactericides

Bioreactors

A bioreactor is an enclosed and sterilized vessel that maintains optimal conditions for the growth of a micro-organism which undergoes fermentation.

Products, such as ethanol or biogas, can be collected from a fermenter after a fixed length of time (batch cultivation) or ongoing (continuous cultivation). In batch cultivation, the micro-organism goes through all the stages of growth prior to the collection of product. In continuous cultivation, the micro-organism is maintained at a peak rate of growth (exponential phase).

Probes and sensors are used to monitor conditions within the fermenter in order to maintain optimal levels of microbial growth. Motorized stirring paddles distribute heat and dissolved substances evenly within the reaction chamber. An external water jacket can be used to absorb excess heat and maintain a constant optimum temperature.

An aerator can introduce compressed air (oxygen) into the chamber, while a defoamer can stop the formation of foam. Acid/base inlets allow for the regulation of pH levels within the chamber (the formation of products, such as carboxylic acids, will alter pH). Nutrient inlets and exhaust outlets allow for the introduction of sugars or amino acids or the removal of metabolic wastes.

Figure B.5 shows how penicillin can be produced on a large scale in a bioreactor.

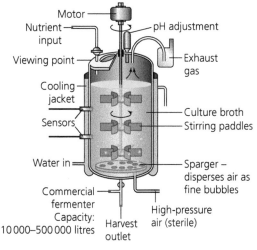

Issues for the commercial fermenter:

1 Can aseptic conditions be maintained?
 – effective sterilization procedures with large quantities

2 Are the culture broth and micro-organisms adequately mixed in bulk?
 – effectiveness of mixing paddles

3 Can the heat generated by the micro-organisms be dispersed?
 – design and operation of the cooling jacket

4 Can aerobic conditions be maintained throughout the broth?
 – sparger design and pressure of air supply to deliver small bubbles

5 Is the anti-foam agent effective?
 – prevention of excess foam at the top of the fermenter

6 Is the pH adequately detected and maintained?
 – early detection of change and its immediate correct adjustment

7 Do the micro-organisms function as anticipated so that yield is maintained?
 – regular sampling of the broth for build-up of antibiotic concentration

Figure B.5 The large-scale production of the antibiotic penicillin

Glossary

Abiotic – the non-living components of an ecosystem.

Abiotic factor – a non-living, physical factor that can influence an organism or ecosystem – for example, temperature, sunlight, pH, salinity or precipitation.

Accuracy – how close to the true value a result is.

Active site – region of enzyme molecule where the substrate molecule binds and catalysis occurs.

Analysis/analyse – break down in order to bring out the essential elements or structure.

Aseptic techniques – procedures to prevent contamination from pathogens. These can include use of disinfectants and using a flame to kill micro-organisms.

Behaviour – the way in which organisms respond to the environment and to other members of the same species.

Biochemistry – the study of the chemical compounds and processes occurring in organisms.

Biomass – the mass of organic material in organisms or ecosystems, usually stated per unit area.

Biotic – the living part of an ecosystem (the community).

Biotic factor – a living part of an ecosystem, such as a species, population or community that influences an ecosystem.

BLAST – an acronym for Basic Local Alignment Search Tool.

Catalyst – substance that speeds up the rate of a chemical reaction. Catalysts are effective in small amounts and remain chemically unchanged at the end of the reaction.

Cladogram – a diagram used in cladistics which shows evolutionary relationships and speciation among organisms.

Community – a group of populations of different species living and interacting with each other in a habitat.

Confounding variables – any other variable, besides the independent variable, that also has an effect on the dependent variable.

Control – an experiment where the independent variable is either kept constant or removed. This can be used for comparison, to prove that any changes in the dependent variable in experiments when the independent variable is manipulated must be due to the independent variable rather than other factors.

Controlled variable – a variable that is kept the same in an investigation. In an experiment, at least three should be listed, and information about how they will be kept the same included.

Correlation – when one variable changes with another variable, so there is a relation between them. The strength of a correlation can be measured using a correlation coefficient. A correlation need not be a causal relation.

Data – recorded products of observations and measurements.

Denaturation – a conformational change in a protein that results in a loss (usually permanent) of its biological properties.

Dependent variable – the variable that is being measured in an investigation.

Diffusion – movement of particles from higher to lower concentration.

Diversity – a generic term for heterogeneity (that is, variation or variety). The scientific meaning of diversity becomes clear from the context in which it is used; it can refer to heterogeneity of species or habitat, or to genetic heterogeneity.

Diversity index – a numerical measure of species diversity calculated by using both the number of species (species richness) and their relative abundance.

Ecology – the study of living organisms in their environment.

Ecosystem – a community of organisms and the environment in which they live and interact.

Enzyme – usually a protein (a very few are RNA) that functions as a biological catalyst.

Ethology – the study of animal behaviour in natural conditions.

Evaluated/evaluation – make an appraisal by weighing up the strengths and limitations.

Evolution – inheritable changes in a population or species over time.

Explanation/explain – give a detailed account including reasons or causes.

Exposure limit – this is the established concentration of a chemical that most people could be exposed to in a typical day without experiencing adverse effects. Exposure limits help in understanding the relative risks of chemicals.

Gel electrophoresis – a process that uses an electric field to separate proteins or fragments of DNA according to size.

Germination – the resumption of growth by an embryonic plant in a seed or fruit, at the expense of stored nutrients.

Hill reaction – the reduction of an electron acceptor by the hydrogens of water, resulting in the release of oxygen. In plants, the final electron acceptor is NADP$^+$. The rate of the Hill reaction can be measured in isolated chloroplasts exposed to light in the presence of an artificial electron acceptor.

Hypertonic – when the external solution is more concentrated (has a higher solute potential) than the cell solution (cytosol), and there is a net flow of water out of the cell by osmosis.

Hypothesis (*general*) – a tentative explanation of an observed phenomenon or event that can be investigated using the scientific method.

Hypothesis (*statistical test*) – there **is** a statistically significant difference between two variables.

Hypotonic – when the external solution is less concentrated (has a lower solute potential) than the cell solution (cytosol), and there is a net inflow of water into the cell by osmosis.

Independent variable – the variable that is being changed in an investigation.

Innate behaviour – behaviour that is inherited from parents and so develops independently of the environmental context. Innate behaviour includes behaviour that is due to a reflex action.

Investigation – a scientific study consisting of a controlled experiment in the laboratory or field-based studies involving sampling.

Isotonic – when the external solution is the same concentration (has the same solute potential) as the cell solution (cytosol), and there is no net entry or exit of water from the cell by osmosis.

Learned behaviour – behaviour that develops as a result of experience.

Limiting factor – factor present in the environment in such short supply that it restricts life processes; for example, it may be a variable that restricts the rate of photosynthesis.

Measuring/measure – obtain a value for a quantity.

Mesocosm – enclosed experimental area that is set up to explore ecological relationships. Because it is a contained experimental area it can be closely controlled and variables monitored.

Metabolism – all the enzyme-catalyzed reactions in a cell or organism.

Mitotic index – the number of cells undergoing mitosis divided by the total number of cells visible.

Motile organism – an organism that can actively move from place to place.

Non-motile organism – an organism that cannot move or, for the purposes of sampling, can only move very slowly (such as limpets on the rocky shore).

Null hypothesis (*statistical test*) – there is **no** statistically significant difference between two variables.

Osmolarity – the concentration of a solution expressed as the total number of solute particles per litre.

Osmosis – the diffusion of water molecules across a partially permeable membrane, from lower to higher solute concentration. Movement is passive.

Photolysis – chemical decomposition and the splitting of water molecules by the absorption of light energy.

Photosynthesis – the production of carbohydrates in cells using light energy.

Physiology – the study of the physical and chemical processes needed by living organisms to perform all functions and activities.

Population – a group of organisms of the same species that live in the same area at the same time.

Precision – describes the reproducibility of repeated measurements of the same quantity and how close they are to each other. Note, measurements can be precise but not accurate.

Prediction/predict – give an expected result.

Processed variable – a variable that can be produced by transforming a measured variable through mathematical manipulation.

Product – what the substrate is converted into in a reaction, catalyzed by an enzyme.

Quadrat – a square frame which outlines a known area for the purpose of sampling.

Random sampling – a method of choosing a sample from a population without any bias.

Rate of reaction – change in amount of substrate that has been converted from a reaction mixture, or the change in amount of product that has accumulated, in a period of time.

Replicate – a repeating of the entire experiment run at the same time.

Respiration – the controlled release of energy in cells from organic compounds, to produce ATP.

Respiratory quotient (RQ) – ratio of the volume of carbon dioxide produced to the volume of oxygen used in respiration.

Sample – a sub-set of a whole population or habitat used to estimate the values that might have been obtained if every individual or response was measured.

Sample size – the number of samples taken from a population.

Scientific method – the use of controlled observations and measurements during an experiment to test a hypothesis.

Sensitivity – the number of significant digits to which a value can be reliably measured. For example, if a digital thermometer can measure to two decimal places, this is the sensitivity of data that can be recorded.

Species – groups of organisms that can potentially interbreed to produce fertile offspring.

Standard deviation – the spread of a set of data from the mean of the sample; it is a measure of the variability of a population from a sample. A small standard deviation indicates that the data are more reliable.

Standard error – an estimate of the reliability of the mean of a population sample. A small standard error indicates that the mean value is close to the actual mean of the population.

Statistical significance – a calculated value that is used to establish the probability that an observed trend or difference represents a true difference that is not due to chance alone.

Substrate – a molecule that is the starting reactant for a biochemical reaction and that forms a complex with the active site of a specific enzyme.

Succession – the process of change over time in a community.

Tidal volume – the volume of air that a human breathes into and out of their lungs while at rest; this is around 500 cm^3, on average.

Transect – arbitrary line through a habitat, selected to sample the community.

Transpiration – the evaporation of water from the spongy mesophyll tissue and its subsequent diffusion through the stomata.

Treatments – well-defined conditions applied to the sample units.

Variable – a factor that is being changed, measured, or kept the same in an investigation.

Ventilation rate – number of breaths per minute.

Index

Answers

Chapter 1

Page 5

1 1 nm = 10^{-3} μm, so 0.133 nm = 1.33×10^{-4} μm = 133 pm = 1.33 Å

Page 6

2 1.002×10^3, 5.4×10^1, 6.9263×10^9, -3.93×10^2, 3.61×10^{-3} and -3.8×10^{-3}

3 1930, 30.52, -429, 6 261 000 and 0.000 000 095 13

4 a nanogram, ng
   b microsecond, μs
   c millimetre, mm
5 a 10^{-12} second
   b 4.0 km
   c 4.56×10^{-3} g

Page 9

6 5.42×10^7 is 10 orders of magnitude larger than 4.70×10^{-3}.

7 The ribosome is 100× smaller than the length of an *Escherichia coli* bacterium (2 μm = 2000 nm; 2000/20 = 100); this is equivalent to 2 orders of magnitude.

8 The number 14.44 has four significant figures – all non-zero digits are significant. In the number 9 000, since there is no decimal point, the zeros may or may not be significant. When considering numbers with zeros at the end we must state the number of significant figures. The number 3 000.0 has five significant figures – the decimal point implies that we have measured to the nearest 0.1. The number 1.046 has four significant figures – zeros between digits are significant. The number 0.26 has two significant figures – zeros to the left of the decimal point only fix the position of the decimal point. They are not significant. The rules are the same when dealing with numbers expressed in standard form, so although 6×10^{23} has one significant figure, 6.02×10^{23} has three significant figures.

Page 10

9 654.389 becomes 654 because the first non-significant digit is 3. 65.4389 becomes 65.4. 654 389 becomes 654×10^3 because we need to put the zeros in to hold the place values. 56.7688 becomes 56.8 because the first non-significant digit is 6. 0.035 422 10 becomes 0.0354. Note that three significant figures is not the same as three decimal places, which would give 0.035.

Page 13

10 5.0 s, 51.2 cm, 20.0 missing units, speed is 0.01 m s^{-1}; mass of the beetle is 4.25 g

Chapter 2

Page 25

1 Enzymes are highly specific in their action, because of the way they bind with their substrate at the active site (three-dimensional pocket or crevice in the protein). At the active site, the arrangement of a few amino acid molecules in the protein (enzyme) matches certain groupings on the substrate molecule,

enabling the enzyme–substrate complex to form; as this complex forms, it seems a slight change of shape is induced in the enzyme molecule. It is this change in shape that is important in raising the substrate molecule to the activated conformation in which it is able to react. Meanwhile, other amino acids of the active site bring about the specific catalytic reaction mechanism, perhaps breaking particular bonds in the substrate molecule and forming others. Different enzymes have different arrangements of amino acids in their active sites; consequently, each enzyme catalyzes either a single chemical reaction or a group of closely related reactions.

2 a If substrate is present in excess, increasing the substrate concentration further will have no effect on the rate of reaction (saturation kinetics).

   b The rate of reaction will increase.

3

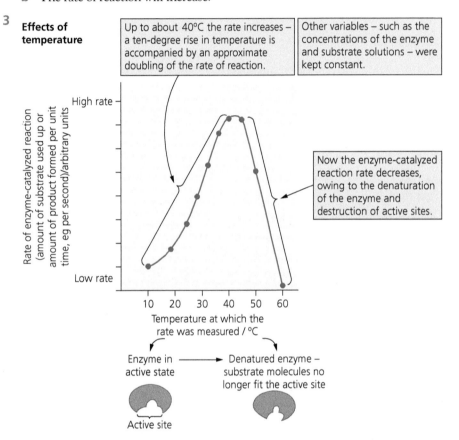

Temperature and the rate of an enzyme-catalyzed reaction

From 10 to 40 °C, increasing temperature increases the average kinetic energy of enzyme and substrate; there are increased collisions per unit time between enzyme and substrate, so more enzyme–substrate complexes form per unit time and the rate of reaction increases. The optimum temperature is at 40 °C, where the rate of enzyme activity is at its peak. Above 40 °C, the rate of reaction decreases as the enzyme denatures: thermal energy disrupts the hydrogen bonds holding the enzyme together. As a result, the active site changes shape, so the substrate can no longer fit in the active site. The enzyme cannot bind with the substrate and behave as a catalyst.

Effects of pH

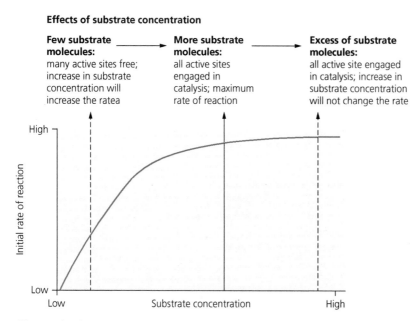

pH at which the
rate was measured

Lower pH	Optimum pH for enzyme	Higher pH

Substrate molecules no
longer fit the active site

Enzyme in
active state

Substrate molecules no
longer fit the active site

Active site

Structure of protein changes when a change of pH
alters the ionic charge on $-COO^-$ (acidic) and $-NH_3^+$ (basic)
groups in the peptide chain, so the shape of the active site is lost

Effects of pH on enzyme shape and activity

Changing the pH will alter the charge of the enzyme due to ionization, which
may change the shape of the molecule so the shape of the active site is lost.
Changing the shape or charge of the active site will diminish its ability to
bind with the substrate, halting enzyme function. Enzymes have an optimum
pH and moving outside this range will result in a diminished rate of reaction;
different enzymes may have a different optimum pH range. A drop in activity
at low and high pH values is mainly because groups on the enzyme donate
or take up protons as part of the catalytic mechanism (general acid and
general base catalysis), and cannot do this at pH extremes; the protonation
state needed for catalysis of enzymes is therefore affected at non-optimum pH
values.

Effects of substrate concentration

**Few substrate
molecules:**
many active sites free;
increase in substrate
concentration will
increase the ratea

**More substrate
molecules:**
all active sites
engaged in
catalysis; maximum
rate of reaction

**Excess of substrate
molecules:**
all active site engaged
in catalysis; increase in
substrate concentration
will not change the rate

High

Initial rate of reaction

Low

Low	Substrate concentration	High

Effects of substrate concentration on enzyme activity

4 Denaturation is a structural change in a protein that alters its three-dimensional shape; the change in shape of the active site means that the substrate can no longer bind to form an enzyme–substrate complex. Denaturation occurs when the bonds and weaker intermolecular forces within a globular protein, formed between different amino acid residues, break, changing the shape of the active site; no peptide bonds are broken during denaturation. Exposure to heat causes atoms to vibrate violently and this disrupts bonds and intermolecular forces within globular proteins; heat causes irreversible denaturation of globular proteins. Small changes in the pH of the medium similarly alter the shape of globular proteins but the structure of an enzyme may spontaneously reform when the optimum pH is restored. Exposure to strong acids or alkalis is usually found to irreversibly denature enzymes.

5 Solutions of hydrogen peroxide are often sold according to their 'volume strength'. The volume strength of a solution of aqueous hydrogen peroxide is measured by the number of volumes of oxygen released when it is completely decomposed under standard conditions (0 °C, 1 atmosphere pressure).

   For example, 1 cm^3 of 20 volume strength hydrogen peroxide solution will release 20 cm^3 of oxygen gas when completely decomposed. Volume strengths can be converted to other measures of concentration, for example, molarity (mol dm^{-3}) or percentage by volume.

6 $\dfrac{1.5}{0.5} \times \dfrac{1.025}{0.025} = 3 \times 41 = 123$

7 The control for this experiment could consist of a soybean solution in distilled water of a specific concentration, maintained at a specific temperature and specific pH for a specific time period (before analysis).

   The result of this control is expected to be no detectable free amino acids, since the enzyme is absent and is assumed to be solely responsible for the digestion of the protein in solution. The soybean protein should not undergo spontaneous decomposition on exposure to light, water (hydrolysis) and/or air (oxygen).

   The comparison consists of a test (in which the numeric results are either zero or have been averaged) whose results are used as a comparison with the results of other tests. For example, in an investigation of soybean protein digestion by the action of papain, you could compare the results of the control with the results of a test reaction between an 80 mg dm^{-3} soybean protein solution and a 10 mg dm^{-3} papain solution maintained at room temperature for 60 minutes at pH 8.00.

   The results of other tests, involving changes in papain solution concentration, papain solution temperature, soybean protein concentration, pH, change in substrate or change in enzyme, etc. could all be compared with the mean results of the control and/or the comparison to see how much faster or slower the new reactions are.

Page 28

8 Pre-incubation is required to ensure that when the reactants are mixed, the reaction occurs at a known, pre-selected temperature.

Page 33

9 Colorimeters can be used to determine the approximate concentration of a highly coloured substance in solution. Some substances, such as chlorophylls, beta-carotene, hemoglobin, myoglobin, anthocyanins (plant pigments found in berries) and betaine (from beetroot), are strongly coloured; with other substances, a colour is produced by adding an appropriate reagent. Colourless compounds that absorb in the ultra-violet region, such as polyphenols and colourless proteins, can be analyzed using a spectrophotometer.

   For example, the concentration of starch in aqueous solution can be determined by adding iodine solution and using a colorimeter to measure the resulting blue-black of the starch–iodine complex. Benedict's test can be made quantitative using a colorimeter, but the precipitate of copper(I) oxide has to be removed by filtration or centrifugation.

A simpler method is to use a solution of 5-dinitrosalicyate (DNSA) which gives more sensitive results than Benedict's test. When boiled with reducing sugars, DNSA changes from yellow to red. The darker the colour with DNSA, the greater the concentration of reducing sugar present in the test substance. A colorimeter can also be used to make the biuret test quantitative.

Determination of the concentration of a substance requires the use of a calibration line. This is prepared by plotting a graph of absorbance of a range of standard solutions against their concentrations. The concentration of the test solution is then determined (via interpolation) by reading a value from the calibration line (see figure below).

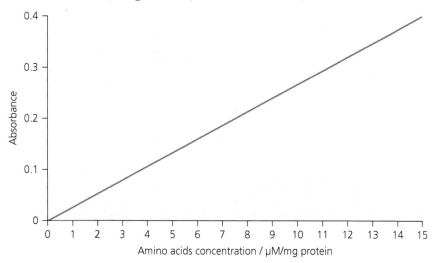

Calibration line for determination of amino acids in solution

Page 36

10 $V_1 = V_2 \times \left[\dfrac{c_2}{c_1}\right] = 100\,\text{cm}^3 \times \left[\dfrac{15\,\text{mM}}{100\,\text{mM}}\right] = 15\,\text{cm}^3$. Thus, add $15\,\text{cm}^3$ of $100\,\text{mM}$ ATP stock to $85\,\text{cm}^3$ of distilled water.

Page 37

11 $50\,\text{mM} = 5 \times 10^{-3}\,\text{M}$; dilution factor = 5; final dilution = $1 \times 10^{-3}\,\text{M}$

Chapter 3

Page 46

1 Biological material varies hugely in size and so it is important to know how large an object is when drawn from a microscope. Magnification gives an indication of the actual size of an object, so that accurate relative comparisons can be made to other objects.

Magnification figures on printed pictures can be misleading because images are sometimes resized to fit the page, making any magnification number provided in the figure legend incorrect. A scale bar can be used to make accurate measurements on a picture: when a picture is resized the bar will be resized in proportion and so the actual magnification can easily be calculated.

2 Use a sharp pencil and draw clear, continuous lines. Do not use any form of shading. It may be helpful to use a magnifying glass and illumination (light). Include a title (with binomial name) and a scale.

If you are drawing from a microscope, state the combined magnification of the eyepiece plus objective lenses used when making the drawing, eg ×100 (low power (see figure below)) or ×400 (high power). Note that this is not the same as recording the scale.

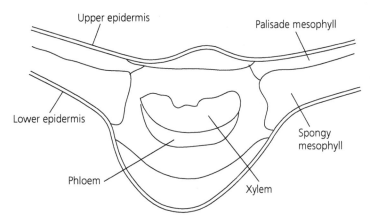

Upper epidermis

Palisade mesophyll

Lower epidermis

Spongy mesophyll

Phloem

Xylem

A low-power drawing of a cross-section of a leaf of privet, *L. ovalifolium*

Label the drawing. Use a sharp pencil and label all relevant structures, including tissues if a microscope is being used to view the specimen. Use a ruler to draw label lines and scale bars. Label lines should start exactly at the structure being labelled. Arrange label lines neatly and make sure they do not cross over each other. Labels should be written horizontally and not written at the same angle as the label line. Add a scale bar immediately below the drawing.

Annotate the drawing. Annotation adds concise notes about the structures labelled on a biological drawing. It is often used to draw attention to structural or functional features of particular biological interest.

▪ Page 50

3 Number of cells in photomicrograph = 100, 11 of which are in mitosis. Mitotic index = $\frac{11}{100}$ = 0.11 (or, expressed as a percentage, 11 % of the cells are in mitosis)

Chapter 4

▪ Page 56

1

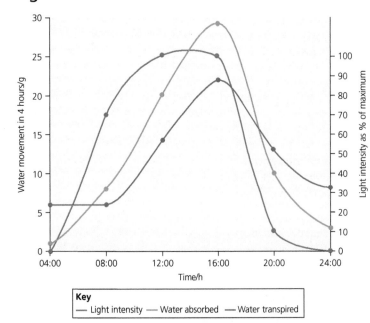

2 At night (24:00 and 04:00), water absorption exceeds transpiration. During the day, more water is transpired than absorbed. The highest levels of water absorption and transpiration occur at 16:00. In this investigation, mass of water transpired (71 g) is approximately equal to mass of water absorbed (69 g).

3 The water that evaporates from the walls of the mesophyll cells of the leaf is continuously replaced. Xylem vessels are full of water. Water moves up the xylem in columns, held together by two different forces: cohesive forces and

adhesive forces. Cohesive forces are due to hydrogen bonds; water molecules stick together because they are polar and so are strongly attracted to each other. Adhesive forces cause water molecules to bind weakly to the sides of the xylem vessels. As a result of these two forces, the column of water is maintained as it is pulled up the plant. The pull of water up the plant and the force of gravity cause tension in the water column. This tension is transmitted down the stem to the roots because of the cohesion of water molecules; at the same time, adhesive forces help to stop the column of water breaking. Consequently, under tension, the water column does not break or pull away from the sides of the xylem vessels. The result is that water is pulled up the stem. Water flow in the xylem is always upwards, and is called the transpiration stream.

Water moves into the roots of a plant by osmosis, replacing water lost by transpiration. Approximately 5–10 % of water entering the plant is used for photosynthesis and maintaining turgor pressure; the rest is lost by transpiration.

Water uptake occurs from the soil solution that is in contact with the root hairs. There are three possible routes of water movement through plant cells and tissues:

- the apoplast pathway: Mass flow occurs through the interconnecting free spaces between the cellulose fibres of the plant cell walls. This pathway passes water through the non-living parts of the cell and the inter-cell spaces, avoiding the living contents of cells.

- the symplast pathway: Diffusion occurs through the cytoplasm of cells and via the plasmodesmata. The plant cells are packed with many organelles which offer resistance to the flow of water, so this pathway is very significant.

- the vacuolar pathway: osmosis occurs between the vacuoles of cells, driven by a gradient in osmotic pressure.

Active uptake of mineral ions in the roots causes absorption of water by osmosis.

There is therefore a continual movement of water up the plant, supported by hydrogen bonds between water molecules. As water moves up xylem vessels, further water molecules enter the roots by osmosis, move into the xylem, and are pulled up the plant.

Page 57

4   A: guard cell; B: upper epidermal cell; C: phloem

5   Allows leaf to float on water

6   Allows leaf to float on water to increase exposure to light for photosynthesis

7   In this leaf, most guard cells are found on the upper epidermis, while in land plants, most guard cells are found on the lower epidermis. In this plant, air chambers are found in both the palisade mesophyll layer and the spongy mesophyll layer, while in land plants, air chambers are found mainly in the spongy mesophyll layer. In this plant, chloroplasts are found in the lower epidermal cells, while in land plants, chloroplasts are not found in lower epidermal cells. In this plant, there is no cuticle on the lower surface, while in land plants, the cuticle is found on the lower surface. [any 3]

Page 60

8   In the respirometer, the far side of the U-tube manometer is the control tube (A). Here, conditions are identical to those in the respirometer tube, but no living material is present. However, any change in external temperature or pressure is experienced equally by both tubes: the effects on the level of manometric fluid are equal and opposite so they cancel out.

9   Soda lime removes the carbon dioxide produced by the respiring maggots so, in the first experiment, the change in volume ($30 \, mm^3 \, h^{-1}$) is due to the oxygen used inside the system. When water is used instead of the soda

lime, the CO_2 released by the animals replaces the used volume of O_2. The difference in volume between this experiment and the one before is the volume of carbon dioxide produced by the respiring maggots, that is, $27\,mm^3\,h^{-1}$.

10 Your evaluation should consider the following questions: Is it acceptable to remove animals from their natural habitat for use in an experiment? Can the animals be safely returned to their habitat? Will the animals suffer pain or any other harm during the experiment? Can the risk of accidents that may cause pain or suffering to the animals be minimized during the experiment? Can contact with the alkali be prevented? Is the use of animals in the experiment essential or is there an alternative method that avoids using animals (for example, using plant material)?

11
Environmental variables	Plant or leaf variables
Light intensity (level of brightness)	Leaf colour (chlorophyll concentration)
Colour (energy/frequency/wavelength) of light	Leaf size
Temperature	Stomata density
Hydrogencarbonate ion concentration	Stomata distribution
Direction of incoming light	Light starved leaves versus leaves kept in bright light
pH of solution	Type of plant
	Leaf age
	Leaf variegation
	Respiration level

▧ Page 64

12 Provides dissolved carbon dioxide, a reactant for photosynthesis

13 Oxygen gas is produced by photosynthesis and trapped in intracellular air spaces, reducing the density of the leaf discs

14 The shorter the time taken, the faster the average rate of photosynthesis

15

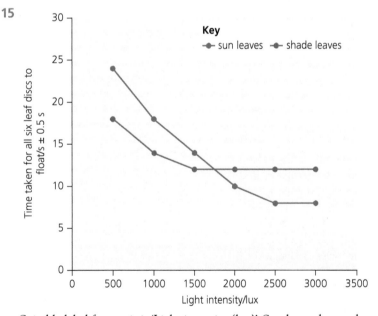

Suitable label for x-axis is 'Light intensity (lux)'. Sun leaves have a lower rate of photosynthesis at low light intensities, compared with shade leaves; sun leaves have a higher rate of photosynthesis at high light intensities, compared with shade leaves. The maximum rate of photosynthesis is lower for shade leaves.

16 The tree can photosynthesize for longer periods (in the early morning and late evening when light intensity is low and in the afternoon when light intensity is high); this allows it to maximize the amount of sunlight used for photosynthesis (shade leaves at the bottom of the canopy use any sunlight that passes through and is not used by the top canopy).

■ Page 66

17 Between 10 am and 4 pm, the dry mass of leaf discs increases because photosynthesis occurs / there is an increase in the rate of photosynthesis. This leads to the production of more glucose which will eventually be stored as starch.

Between 4 pm and 10 pm, the dry mass of leaf discs decreases because respiration occurs in the plant / photosynthesis has stopped. During respiration, starch is converted to glucose which will be broken down (hydrolysed).

■ Page 70

18 Dark blue colour of dye disappears as it is reduced

19 Tube 2 – shows that light alone cannot induce colour change; tube 3 – shows that light is needed

20 a Nuclei (slightly larger) / mitochondria (approximately the same size / slightly smaller)
   b These organelles lack chlorophyll.

21 To reduce enzyme activity / so that enzymes of lysosomes do not digest cell contents.

22 Cell reactions need a certain pH; acids can be released when cells are broken

23 a DCPIP
   b H_2O

Chapter 5

■ Page 72

1 For example: temperature, measured using a thermometer – put thermometer at set distance below water; light, measured using a light-meter – put light-meter at set distance above or below water; flow velocity, measured using a flow-meter – put propeller at set distance below water and count revolutions per minute; dissolved oxygen, measured using an oxygen-meter. For each measurement, repeat several times to improve reliability.

2 For example: light – would decrease with depth; turbidity – would increase with depth; temperature – would decrease with depth.

3 For example: soil moisture – if soil is too hot when evaporating water, the organic content can also burn off.

■ Page 74

4 The capture–mark–release–recapture method is used to determine the population size of animals in large areas. In this method, a sample of organisms is captured, marked and released. Then, after a suitable time interval, a second sample of animals is captured and the number of marked individuals noted. The population size (N) is calculated by multiplying the number of animals captured (marked and released) on the first day (n_1) by the number of animals captured on the second day (n_2), divided by the number of marked animals

recaptured on the second day (m): $N = \dfrac{n_1 \times n_2}{m}$.

When using this method:

■ there must be an adequate time interval between the capture and recapture of animals

■ the marking method must not decrease the chances of survival of the organisms / must not harm the animals / must not introduce bias when recapturing animals

This method can only be used to calculate the population size when there is enough / relevant data. This method can be used to calculate the population of organisms such as snails / dragonflies / small mammals such as mice / *other suitable examples.*

Page 76

5 For example:

Light: A light-meter can be used to measure the light intensity in an ecosystem; the meter should be held at a standard, fixed height above the ground and read when the value is steady and not fluctuating; cloud cover and changes in light intensity during the day mean that values must be taken at the same time of day and same atmospheric conditions: this can be difficult if several repeats are taken; the direction of the light-meter also needs to be standardized so it points in the same direction at the same angle each time it is used; care must be taken not to shade the light-meter during a reading.

Temperature: An electronic thermometer with probes (datalogger) allows temperature to be accurately measured in air, water, and at different depths in soil; the temperature needs to be taken at a standard depth; problems arise if the thermometer is not buried deeply enough: the depth needs to be checked each time it is used; temperature can only be measured for a short period of time using conventional digital thermometers: dataloggers can be used to measure temperature over long periods of time and take fluctuations in temperature into account.

Wind: Precise measurements of wind speed are made with a digital anemometer: the device can be mounted or hand-held; the anemometer must be held at a fixed and unvarying height above the ground for each measurement and held so that it is directed into the wind; care must be taken not to block the wind; gusty conditions may lead to large variations in data.

Flow velocity: Surface flow velocity in streams can be measured using a flow-meter (a calibrated propeller attached to a pole); the impeller is inserted into water just below the surface and pointed into the direction of flow; number of readings are taken to ensure accuracy; as velocity varies with distance from the surface, readings must be taken at the same depth; results can be misleading if only one part of a stream is measured; water flows can vary over time because of rainfall or glacial melting events.

Dissolved oxygen: Oxygen-sensitive electrodes connected to a meter can be used to measure dissolved oxygen; readings may be affected by oxygen in the air, so care must be taken when using an oxygen meter to avoid contamination with oxygen in the air.

6 Percentage frequency is the percentage of quadrats in an area in which at least one individual of the species is found. Percentage cover is the proportion of a quadrat covered by a species, measured as a percentage.

7 A sample population needs to be captured, marked, released and recaptured. The formula for the Lincoln index is:

$$N = \frac{n_1 \times n_2}{m}$$

where: N = total population size of animals in the study site; n_1 = total number of animals captured and marked on first day; n_2 = number of animals recaptured on second day; m = number of marked animals recaptured on second day.

8 All biomass from sample area (eg 1 m^2) is removed; with plant material, roots are dug up and removed as well as above-ground biomass. The sample is weighed in a container of known weight, and then put in a hot oven (80 °C). After a specific length of time, the sample is removed from the oven, reweighed and then put back in the oven; this is repeated until the same mass is recorded from two successive readings. No further loss in mass indicates that water is no longer present.

Page 77

9 The habitat with $D = 1.83$ is simpler/younger than the habitat with $D = 3.65$, which could be in a more complex/mature ecosystem. The habitat with the lower D value could have been disturbed – for example, by logging, pollution, colonization, or agricultural management.

■ Page 78

10 Level of isolation from other urban habitats; level of fragmentation; management practices, for example grass cutting, pruning of trees, weed removal; habitat type; habitat age; habitat diversity; habitat connectivity; mobility of animals in the habitat.

■ Page 79

11 A transect is a line through a habitat or environmental gradient, selected to sample the community. Transects are used to measure changes along a gradient; this ensures that all parts of the gradient are measured. A quadrat is a sampling area enclosed within a frame. Quadrats can be used to sample at regular intervals along a transect and are used to estimate the abundance of plants and non-motile animals.

12 (a)

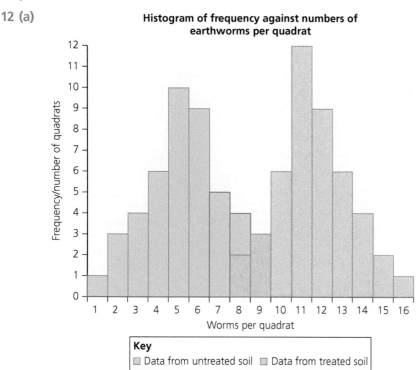

Histogram of frequency against numbers of earthworms per quadrat

Key
Data from untreated soil ▨ Data from treated soil

An investigation into the effect on earthworm populations of pesticide treatment

(b) Untreated soil has a higher abundance of earthworms than soil treated with pesticides. This suggests that pesticides have reduced the abundance of earthworms. This may be because pesticides are non-specific and have affected earthworm populations as well as target pest species: pesticides may decrease the reproductive potential of earthworms / decrease metabolic processes / decrease enzymatic activities / increase individual mortality / decrease growth / change feeding behaviour.

■ Page 80

13 Ecological gradients are found where two ecosystems meet (for example, on beaches or on lake shores) or where an ecosystem suddenly ends (for example, at forest edges). Both biotic and abiotic factors vary with distance and form gradients.

14 Set up a transect and measure abiotic factors at regular intervals (for example, every 5 m) using a standard method.

15 Frame quadrats, which are empty frames of known area (for example, 1 m²); grid quadrats, which are frames divided into 100 small squares; point quadrats, which are made from a frame with 10 holes, inserted into the ground by a leg.

16 A cross staff is used to move a set distance (for example, 0.6 m) vertically up the transect. The staff is set vertically and a point measured horizontally from an eyesight 0.6 m from the base of the staff. Biotic and abiotic factors are measured at each height interval.

Chapter 10

▨ Page 128

1

▨ Histograms and bar charts

In histograms, the groups are shown as intervals along the *x*-axis and there is no 'overlap' between groups so any value can only belong to one group. The number of values in each group, or frequency, is shown on the *y*-axis. This type of graph is often known as a frequency histogram.

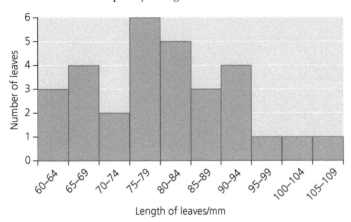

Histogram of leaf length data for holly leaves

Bar charts are drawn where the independent variable is made up of a number of different, discrete categories and the dependent variable is continuous. The blocks can be arranged in any order, but it can help with comparisons if they are arranged in descending/ascending order of size.

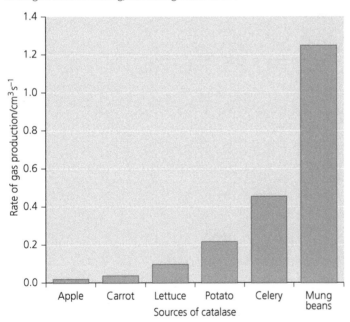

Bar chart showing catalase activity in different plant tissues

▨ Pie charts

Pie charts are circular graphs used to plot proportional data, such as the proportion of cells at different stages of mitosis during cell division. The circle represents the whole and is divided into sectors, each of which is proportional to the size of the sample. The sample values are converted to percentage figures and the size of each sector is determined by calculating the angle that will correspond to the percentage. A pie chart is normally drawn with the sectors in rank order, starting at the 12 o'clock position and moving in a clockwise direction.

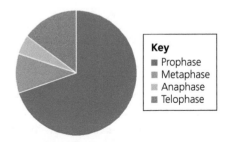

An example of a pie chart, showing the proportion of dividing cells at each stage of mitosis

Box-and-whisker plots

For data tables with non-parametric data the appropriate descriptive statistics are medians and quartiles, and the appropriate graph is a box-and-whisker plot. The ticks at the top and bottom of the vertical lines show the highest and lowest values in the set of data. The top of the box shows the upper quartile, the bottom of the box shows the lower quartile, and the horizontal line within the box represents the median. The central rectangle spans the interquartile range, or IQR).

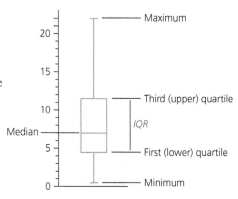

Box-and-whisker plot showing distribution of length data for a selection of maple leaves, indicating minimum, first quartile, median, third quartile and maximum values.

Triangular graphs

These are graphs with three axes which make up an equilateral triangle. They can be used to plot raw data that can be easily divided into three proportions. They are useful when raw data from several sources are plotted on a single graph. An example of such data is provided by the constituent particles of a soil sample:

Particle	Percentage
Silt	5
Clay	33
Sand	62
	100

These values have been plotted on the triangular graph below and the dotted lines show how the values are carried across the graph to meet at one point. The position of this point indicates the relative dominance of sand in this particular soil sample. Samples taken from other locations could be analysed for these particles and plotted on the graph.

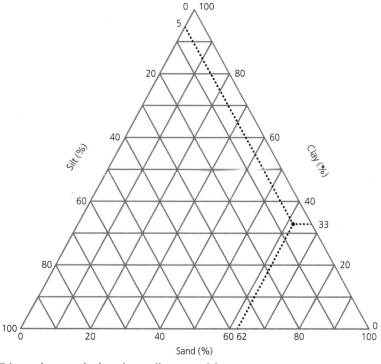

Triangular graph showing soil composition

Kite diagrams

If an area is being studied to determine the changes in the distribution of certain animal or plant species along a belt transect, the data collected can be plotted as a kite diagram. A kite diagram consists of kites drawn along a baseline. The kites represent the abundances of a species; the wider the kite, the more frequent (abundant) the species is.

> **Examiner guidance**
>
> If the kite converges into a line then the species is absent at that point on the transect.

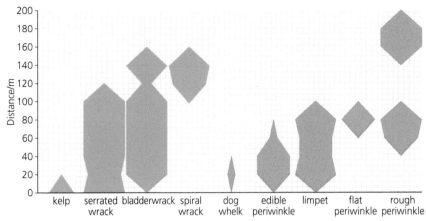

The distribution and abundance of organisms along a belt transect, shown in a kite diagram

ACFOR scale

The actual or relative abundance of a particular species at a particular sample point can be determined by reading the width of the kite from the kite diagram. Kite diagram abundances are sometimes presented using a relative abundance scale known as the ACFOR scale (see table). The qualitative letter scale is then converted to a numeric scale.

ACFOR scale	Abundance scale
Species absent	0
Rare	1
Occasional	2
Frequent	3
Common	4
Abundant	5

Option A

Page 146

1 The behaviour shown by the data on woodlouse activity over a 24-hour period directly favours their chances of survival because the woodlouse is less active during the daylight hours, so avoids many predators. The woodlouse can remain in shaded damp or humid conditions during the day to reduce the probability that it will become desiccated or encounter too much heat. When it is dark however, there are greatly increased numbers of active woodlice. They come out to find food when there are fewer predators around. In addition there is increased reproductive potential as there is a greater probability of woodlice meeting if they are active during the same time period.

Option B

Page 149

1 Sterilization means the complete destruction of all the micro-organisms, including spores, from an object or environment. It is usually achieved by heat or filtration but chemicals or gamma radiation can be used.

Disinfection is the destruction, inhibition or removal of microbes that may cause disease or other problems, eg spoilage. It is usually achieved by the use of chemicals.

2 To make a pour plate, a small sample of inoculum from broth culture is added by pipette to the centre of a Petri dish. Molten, cooled agar medium is then poured into the Petri dish containing the inoculum. The dish is rotated to ensure that the culture and medium are thoroughly mixed and the medium covers the plate evenly.

Pour plates allow micro-organisms to grow both on the surface and within the medium. Most of the colonies grow within the medium; they are small in size and may be confluent. The few colonies that grow on the surface are of the same size and appearance as those on a streak plate.

If the dilution and volume of the inoculum, usually 0.10 cm^3, are known, the viable count of the sample, ie the number of bacteria or clumps of bacteria, per cm^3 can be determined. The dilutions chosen should produce between 20 and 100 countable colonies.

Page 151

3 Batericides are contained in the arms of the mast ring so that sensitivity to many bactericides may be tested simultaneously. There is evidence that growth of the bacterium is more sensitive to certain antibiotics (eg CM, A) than to others (eg S, I).